Solution of Large Networks
by Matrix Methods

SOLUTION OF LARGE NETWORKS BY MATRIX METHODS

HOMER E. BROWN

Eletrobrás Professor of
Power Engineering
Escola Federal de
Engenharia de Itajubá
Brazil

A WILEY-INTERSCIENCE PUBLICATION

John Wiley & Sons, Inc.

New York · London · Sydney · Toronto

Library of Congress Cataloging in Publication Data:

Brown, Homer E   1909–
    Solution of large networks by matrix methods.

    "A Wiley-Interscience publication."

    Includes bibliographical references and index.

    1. Electronic data processing—Electric networks.
2. Short circuits.   3.   Matrices.   4.   Electric engineering—
Mathematics.   I. Title.

TK3226.B763   621.319′2′01512943   74-34159
ISBN 0-471-11045-0

Printed in the United States of America

10 9 8 7 6 5 4 3 2 1

To Mary Isabel

Preface

This book covers the class material that was given to graduate classes in network analysis at Iowa State University, Rensselaer Polytechnic Institute, Purdue University, and The Escola Federal de Engenharia de Itajubá (Brazil) when I was a visiting professor at those institutions. The material has been expanded into book form and is intended in graduate study work to indicate the methods now used in industry. It will also be helpful for practicing engineers who completed their formal education prior to the computer revolution. The methods discussed are illustrated by simple numerical examples for a better understanding of the techniques.

Although electric power systems are subjected to short circuits only a small proportion of the total time, short circuits on networks are treated first in the book, because this subject is far simpler to explain and comprehend than is the solution of normal power flow problems. Because transient stability is more complex than power flow problems, this subject follows.

Finally an introduction is given to optimization methods such as linear programming, steepest ascent, and eigenvalues since the next development in network analysis will surely exploit these techniques.

I should like to thank every individual that helped me in writing the book but this would be impossible because of the great number. Therefore, I list only a few names for special commendation. Of the many former associates at The Commonwealth Edison Company, I express my gratitude to Conrad E. Person and Robert G. Andertich who assisted in developing some of the techniques that are discussed in the text. For my first opportunity to be a visiting professor on a university campus while on loan and financially supported by The Commonwealth Edison Company, I am indebted to Professor Paul M. Anderson of Iowa State University and Vice President Ludwig F. Lischer of The Commonwealth Edison Company. I am especially grateful to Dr. Eric T. B. Gross for the opportunity to be a visiting professor three times in his Power Engineering Program at Rensselaer and for his encouragement to begin writing my lectures in book form. To Dr. T. S. Lauber for reviewing the manuscript and suggesting modifications that would improve the clarity of the material, I am greatly indebted. I acknowledge the help of Dean Amadeu Casal Caminha and Professor José Abel Royo dos Santos of the Escola Federal de Engenharia de Itajubá for their corrections in the English text when translating it into Portuguese. I am also indebted to Dean Caminha for furnishing the secretarial help of Lair Elisa Fernandes and Sônia Maria Maia. I commend the women for their great care in typing in a foreign language. Finally the writing of the book would not have been possible without the loving understanding of my wife, Mary Isabel.

HOMER E. BROWN

Itajubá, Minas Gerais, Brazil
September 1974

Contents

1 GENERAL BACKGROUND   **1**

Early Computation Methods   1
The Computer   2
Computational Methods   2
Power Flow Problems   2
Short-Circuit Studies   3
References   5

2 MATRIX ALGEBRA   **6**

Definition of a Matrix   6
Order of a Matrix   6
The Transpose of a Matrix   7
Equal Matrices   7
The Sum of Matrices   7
    Examples   8
Multiplication of a Matrix by a Scalar   8
Multiplication of Matrices   8
    Example   9
The Identity Matrix   10
The Determinant of a Matrix   10
Minors and Cofactors   10
Matrix Inversion   11
Classical Matrix Inversion   12
Shipley's Inversion   14
    Example of a Shipley Inversion   16
    Justification of the Shipley Inversion   18
Partitioning of Matrices   21
Exercises   22
References   22

3 THREE-PHASE SHORT-CIRCUIT CALCULATIONS   **23**
Description of the Z-Bus Matrix   24
The Z-Matrix Building Algorithm   28
Data Preparation   28
A Reference Line to a New Bus   28
Addition of a Radial Line to a New Bus   29
Addition of a Loop Closing Line   30
Kron Reduction of a Matrix   32
An Illustrative Example of the Building Algorithm   34
   Addition of the First Line   35
   Addition of the Second Line   35
   Addition of the Third Line   36
   The Matrix Reduction—A Delta-Star Conversion   38
   Addition of the Fourth Line   38
   Addition of the Fifth Line   39
   Addition of the Sixth Line   40
   Addition of the Seventh Line   40
   Addition of the Eighth Line   41
   Addition of the Ninth Line   41
   Addition of the Tenth Line   42
   Addition of the Eleventh Line   42
   Addition of the Last Line   42
Fault Analysis of a System   43
Voltages on Buses During Fault Condition   43
Opening a Line During a Study   44
Bus Tie Breakers   46
Bus Tie Breaker That Can Be Opened   47
Limitation of the Z-Matrix Method   47
References   48

4 EXTENSION OF THE Z-MATRIX SHORT-CIRCUIT
PROGRAM TO INCLUDE VERY LARGE SYSTEMS   **49**
Data Preparation   50
Details of the Line Reordering Technique   52
Example of the Line Reordering Routine   53
Processing Lines of Infinite Impedance   55
   Loop Closing Lines   55
   A Line from the Reference to a New Bus   56
Modification of the Building Algorithm   58
Reduction of a System to an Equivalent   59
Computation of an Equivalent   59
   Example   60

Extension of the Method to Large Systems   **63**
The Large Short-Circuit Program   **65**
     Example   **66**
Large Equivalents   **69**
Summary   **71**
References   **74**

5 SINGLE-PHASE SHORT CIRCUITS   **75**
The Matrix Method   **75**
System Changes   **79**
An Additional Complication   **79**
Mutual Coupling Compensation   **80**
The Modified Logic   **80**
Modified Radial Line Algorithm   **80**
     For the Case $K \neq Q$ **81**
     For the Case $K = Q$ **83**
Addition of a Loop Closing Mutually Coupled Line   **85**
     The Diagonal Element   **87**
     Example   **88**
Opening a Mutually Coupled Line   **95**
     Exercise   **97**
Flow in Lines in the Zero Sequence Network   **98**
     Example   **98**
References   **100**

6 POWER FLOW SOLUTIONS   **102**
Gauss-Seidel Method   **104**
Newton-Raphson Power Flow   **107**
     When $k \neq m$ **110**
     For the Case $m = k$ **111**
Forming the Jacobian Matrix   **113**
Refinements in the Newton's Method   **118**
The Bus Impedance Matrix Power Flow   **119**
References   **122**

**7 HIGH-SPEED REDUCED ACCURACY POWER FLOW
CALCULATIONS FOR CONTINGENCY EVALUATION AND
MAXIMUM INTERCHANGE CAPABILITY DETERMINATIONS**   **124**
Axis Discarding in Z-Matrix Methods   **125**
The Z-Matrix Contingency Evaluation Method   **129**
A Multiple Contingency   **135**
Contingency Caused by a Line Addition   **138**

Interchange Capability Evaluation (Single)   141
Maximum Interchange Determination   145
The Newton-Raphson Power Flow   159
Review of Some Fundamentals   160
The Decoupling of the Newton-Raphson Equations   162
Development of the Real Power Model   164
Development of the Reactive Power Model   166
Method of Application   174
Comparison of the Z-Matrix and Newton-Raphson Method   175
Distribution Factors   175
References   176

8 TRANSIENT STABILITY   178
Development of the Techniques   180
The Dilemma   181
Classical Representation of a Generator   182
Transient Saliency   185
Excitation Response and Saturation   188
Voltage Regulators   191
Load Flow Calculations   192
    Example 1   193
Transient Stability Study by Coherent Machines   195
Steady-State Stability   196
Network Equivalent with Distribution Factors   198
    Example 2   200
    Example 3   201
References   202

**9 EIGENVALUES, EIGENVECTORS: LINEAR
PROGRAMMING AND OPTIMIZATION**   204
Eigenvalues and Eigenvectors   205
Physical Interpretation of the Problem   206
Quadratic Forms   210
Shifting   214
    Exercise   217
The LR-Transformation   217
A Quadratic Convergent Method   221
Applications   231
Linear Programming   231
The Constraints   232
    Example 1   232
    Example 2   237

Optimization   **242**
The Optimization Problem   **244**
Lagrangian Method   **244**
    Example 1   **245**
Multiple Constraints   **247**
Inequality Constraints   **247**
    Example 2   **248**
The Gradient Method of Optimization   **249**
    Example 3   **250**
References   **252**

INDEX   **254**

General Background

Prior to about 1950 matrices were used only as research tools. They systematized the arrangement of materials and generally forced the research worker to be organized. Matrices at that time in no way reduced the computational effort; however, the absence of high-speed computers limited investigations to small sets of equations involving only very small matrices.

The first generation of small-scale computers extended the use of matrices in solving network problems of limited size [1].

Networks, a broad category of studies, extend into many disciplines. The range of problems includes traffic flow in a network of city streets, stress analysis of the steel framework of large buildings, airplane wings, gas flow in pipes, the flow of electricity in a large electrical network, heat flow, mechanical rotation, and such.

EARLY COMPUTATION METHODS

As recently as 1955 all electrical network problems were solved either by hand or by a "network calculator."* The network calculator was poorly named, since it did not perform any calculations. It is merely an electrical analogue device. For electrical problems the analogue is direct; that is, electrical current in the problem network is represented by current in the analogue, voltage is represented by voltage and so on. The network being investigated is represented by another network on a greatly reduced scale.

The network calculator can also be used to solve problems in other fields, but problem variables must be converted to electrical quantities. For example, the stress and strain in a steel beam could be represented by voltage and current, respectively, while mechanical inertia could be represented by inductance or capacitance.

*Copyright of Westinghouse, GE Co., used "Network Analyser."

THE COMPUTER

The second and third generation of digital computers made possible the investigation of large networks (steel structures, power system networks, etc.) by matrix methods. The superiority of the network calculator as a tool for educating electrical power system operating and system planning personnel is justified, since the network response to the various adjustments (generator angle or voltage level) can be readily observed. However, because of the larger capability of the computer programs and the almost universal availability of the digital computer the computer is superior and more economical for detailed studies of large systems. The network analyzer on the other hand is a very specialized tool, and even during peak usage was not generally available. Consequently, the network calculator has virtually disappeared in the United States.

COMPUTATIONAL METHODS

The availability of the computer changed greatly the mathematical approach used for network solutions. Longhand calculations can be carried out more readily using loop equations. The earliest computer programs analyzing the flow in networks merely automated these longhand methods [2].

Several investigators did considerable work with incident matrix and connection matrix algorithms for automatically determining independent loops in the network, since this was the most difficult part of data preparation for the loop formulation of the problem [3,4]. Later nodal equation metnods were developed and proved to be greatly superior for the computer solution of network problems.

POWER FLOW PROBLEMS

The first computer attempts to solve network flow problems had limited success, because the programs merely automated the longhand methods using loop equations and did not exploit the capability of the computer. The greatest burden in these early programs was the preparation of the data that defined the independent loops of the network. A considerable amount of work was done to develop methods whereby the computer could automate the generation of the loop connection matrix. The method was somewhat successful but in turn added to the burden on the limited computer memory available [3,4].

The first really successful network flow program was developed by Ward and Hale [5]. They used the nodal formulation of the problem and solved the simultaneous quadratic equations that describe the electrical network by a modified Newton iterative procedure [6]. The programs, which followed immediately, implemented the Gauss-Seidel algorithm.

The success of the Ward and Hale method was quickly accepted by the power industry, and a number of papers by Glimn and Stagg, Brown and Tinney, and others described modifications of the algorithm and additional features incorporated in their programs.

The increase in high voltage interconnections between systems in the late 1960s and the availability of large computers greatly enlarged the size of systems studied. Power flow studies of larger systems by the Gauss-Seidel method require a greater number of iterations to obtain a solution or become mathematically unstable, even though the network being studied is actually a workable system. During the iterative process, in the Gauss-Seidel method, the effect of adjustments in an iteration are reflected only to the neighboring nodes. The propagation of an adjustment across a large system therefore takes several iterations. Meanwhile conflicting adjustments may be taking place and are transmitted and reflected across the system.

Fortunately, as early as 1961, researchers were investigating other methods for solving network flow problems. The most successful load flow algorithm and the most universally accepted replacement for the Gauss-Seidel process is the Newton-Raphson method. This algorithm [10] was the result of a continuing development by the Bonneville Power Administration [7–10].

Another load flow method that overcomes the instability of the Gauss-Seidel method is the impedance matrix load flow algorithm [11]. The method has convergence characteristics similar to the Newton Raphson method for the average power systems load flow problems. However, the memory requirements for the impedance matrix are very severe because the Z-matrix is full and not sparce as with the Y-matrix of the Gauss-Seidel or the Jacobian matrix of the Newton-Raphson method. This severe storage problem can be overcome by tearing the system into parts and applying the diakoptics techniques of Kron [12, 13].

SHORT-CIRCUIT STUDIES

The analysis of a network under short-circuit conditions followed a development similar to that described for load flow problems. The earliest attempts were also based on the loop equation formulation of the problem [14]. Again the definition of the loops in the network seriously handicapped the method.

The success of the Ward and Hale iterative method in the load flow problem encouraged research in the use of this method for short-circuit studies [15]. Here an excellent technique was misapplied. In a series of load flow cases the system changes ordinarily alter the voltage profile of the system by a very small amount, and the Gaus-Seidel iterative approach is usually successful. In a short-circuit analysis of a system the voltage profile is drastically changed from one network condition to the next. As each node takes its turn as the node to be short circuited, its voltage is set at zero. Buses that are closely coupled electrically have their voltages greatly reduced from the normal value. The iterative procedure is not well suited to this task. The entire iterative process must be repeated for each fault study condition, whether it is a change in the location of the short circuit or a change in the network configuration, to determine the new system voltage profile. The development of a fast method of assembling the Z-bus (driving point and transfer impedance) matrix eliminated further work on iterative methods for short circuit analysis [16].

If the Z-matrix that describes the network is available (has been computed), the complete fault analysis can be calculated with a minimum of arithmetic operations. The algorithm used to compute the Z-matrix originally can be used to modify the matrix for different system conditions with very little computation. The Z-matrix has enjoyed wide acceptance in the power industry.

The limitation, imposed by the storage of the Z-matrix on system size that can be studied, was overcome by a matrix axis discarding technique described in references 17, 18, and 19.

References

1. **M. Lantz**, Digital short-circuit solution of power system network including mutual impedance, *Trans. AIEE PA&S*, (February 1958), pp. 1230.

2. **J. M. Henderson**, Automatic digital computer solution of load flow studies, *Trans. AIEE PA&S*, (February 1955), pp. 1696.

3. **W. F. Tinney and C. M. McIntyre**, A digital method of obtaining a loop connection matrix, *Trans. AIEE*, Vol. 79, Part III, (1960), pp. 740–746.

4. **H. Edelmann**, Numerical and algebraic generation of mesh-impedance matrices by set-theoretical intersection on a digital computer, *Trans. AIEE*, Vol. 83, Part III, (1964), pp. 397–402.

5. **J. B. Ward and H. W. Hale**, Digital solution of power flow problems, *Trans. AIEE*, Vol. 75, Part III, (1956), pp. 398–404.

6. **Savadore and Baron**, *Numerical Methods in Engineering*, Prentice Hall, 1955, p. 28.

7. **J. E. Van Ness**, Iteration methods for digital load flow studies, *Trans. AIEE PA&S*, Vol. 78A, (August 1959), p. 583.

8. **J. E. Van Ness and J. H. Griffin**, Elimination methods for load flow studies, *Trans. AIEE PA&S*, Vol. 80, (June 1961), p. 299.

9. **N. Sato and W. F. Tinney**, Techniques for exploiting the sparsity of the network admittance matrix, *Trans. IEEE PA&S*, (December 1963), Vol. 82, p. 944.

10. **W. F. Tinney and C. E. Hart**, Power flow solutions by Newton's method, *Trans. IEEE PA&S*, Vol. 86, (November 1967), p. 1449.

11. **H. E. Brown, G. K. Carter, H. H. Happ, and C. E. Person**, Power flow solution by the impedance matrix method, *Trans. IEEE PA&S*, Vol. 65, (April 1963), p. 1.

12. **Gabriel Kron**, Diakoptics—The Piecewise Solution of Large Scale Systems, General Engineering Laboratory Report 57GL 330 General Electric Co., 1957.

13. **R. G. Andertich, H. E. Brown, H. H. Happ, and C. E. Person**, The piecewise solution of the impedance matrix load flow, *Trans. IEEE PA&S*, Vol. 87, (October 1968), pp. 1877–1882.

14. **A. F. Glimn, R. Haberman Jr., J. M. Henderson, and L. K. Kirchmayer**, Digital calculation of network impedances, *Trans. AIEE*, Vol. 74, Part III, (1955), pp. 1285–1296.

15. **L. W. Coombe and D. G. Lewis**, Digital calculation of short-circuit currents in large complex-impedance networks, *Trans. AIEE*, Vol. 75, Part III, (1956), pp. 1394–1397.

16. **H. E. Brown, C. E. Person, L. K. Kirchmayer, and G. W. Stagg**, Digital calculation of three-phase short circuits by matrix method, *Trans. AIEE*, Vol. 79, Part III, (1960), pp. 1277–1281.

17. **H. E. Brown and C. E. Person**, Short Circuit Studies of Large Systems, Power Systems Computer Conference Proc. Vol. 4, Report 4.11 (PSCC), Stockholm, 1966.

18. **H. E. Brown and C. E. Person**, Short Circuit Calculations of Large Systems, *IEEE Power Industry Computer Applications Conference (PICA) Proceedings*, Pittsburgh, 1967, pp. 335–342.

19. **H. E. Brown**, State of the Art of Short Circuit Studies, IEEE Mexico International Conference on Systems, Networks, and Computer, Oaxtepec, Mexico, 1971.

2

Matrix Algebra

For a complete treatment of matrix algebra the reader is referred to a standard textbook on the subject [1]. The matrix algebra given in this book includes only those topics that are required in the later chapters of this book. The main emphasis is on methods suited to treatment of large matrices for the solution of networks using computers.

DEFINITION OF A MATRIX

A matrix is a set of numbers, functions, objects, or operators arranged in a rectangular array of rows and columns, enclosed in brackets, that obeys certain rules for addition and multiplication. The individual items of the array are called elements or scalars, and their location in the array is identified by a double subscript system in which the first subscript indicates the row and the second subscript corresponds to the column. The element a_{ij} is therefore located in the ith row and the jth column of an array. Generally the size of the matrices that are dealt with here are very large. The matrix is then not set down completely, but is represented by a single capital letter or a single element with a double index:

$$Y = [y_{ij}] \qquad i = 1 \text{ to } n, j = 1 \text{ to } m \qquad (2.1)$$

ORDER OF A MATRIX

A matrix that has m rows and n columns is said to be a matrix of order m by n. A matrix with a single row or a single column is referred to as a row vector or a column vector, respectively. When it is completely obvious whether the matrix is a single row or column, it is referred to simply as a

6

vector. The element in a vector generally carries only a single index.

$$I = [i_1 \ i_2 \ i_3 \ i_4] \qquad \text{or} \qquad v = \begin{bmatrix} v_1 \\ v_2 \\ v_3 \\ v_4 \end{bmatrix}$$

A square matrix is a matrix that has an equal number of rows and columns; that is, $m = n$ in equation 2.1.

The major diagonal of a square matrix is the set of elements that appear on the diagonal of the array from upper left to lower right; that is, the elements a_{ii} that have equal indices are diagonal elements.

THE TRANSPOSE OF A MATRIX

The transpose of a matrix A is the matrix designated A' or A_t in which the rows of A are the columns of A'. Thus if $A = [a_{ij}]$ then A' or $A_t = [a_{ji}]$. The transpose of a row vector is a column vector. For example,

$$A = [a \ b \ c \ d] \qquad \text{then} \qquad A_t = \begin{bmatrix} a \\ b \\ c \\ d \end{bmatrix}$$

EQUAL MATRICES

Two matrices are said to be equal only when all the elements of one matrix are identically equal to the corresponding elements of the other matrix; that is, $a_{ij} = b_{ij}$ for every i and j. It is obvious therefore that two matrices that are equal must be of equal order.

THE SUM OF MATRICES

The sum (or difference) of two matrices is defined only for matrices of equal order. Consider two matrices A and B of equal order; their sum,

$C = A + B$ is defined to be the matrix whose elements are $c_{ij} = a_{ij} + b_{ij}$, where the sum of the elements follow the customary algebraic rule of signs. Since any element in a matrix sum is equal to the algebraic sum of the corresponding elements in the matrices being added, the associative and commutative laws for the addition of scalars apply.

$$(A + B) + C = A + (B + C) \qquad \text{associative law}$$

$$A + B = B + A \qquad \text{commutative law}$$

The difference of two matrices $C = A - B$ is defined by $c_{ij} = a_{ij} - b_{ij}$.

Examples The sum

$$\begin{bmatrix} 2 & 0 & -3 \\ 7 & 8 & -1 \end{bmatrix} + \begin{bmatrix} 1 & 2 & 3 \\ 4 & -5 & 6 \end{bmatrix} = \begin{bmatrix} 1 & 2 & 3 \\ 4 & -5 & 6 \end{bmatrix} + \begin{bmatrix} 2 & 0 & -3 \\ 7 & 8 & -1 \end{bmatrix}$$

$$= \begin{bmatrix} 3 & 2 & 0 \\ 11 & 3 & 5 \end{bmatrix}$$

and the difference

$$\begin{bmatrix} 1 & 2 & 3 \\ 4 & -5 & 6 \end{bmatrix} - \begin{bmatrix} 2 & 0 & -3 \\ 7 & 8 & -1 \end{bmatrix} = \begin{bmatrix} -1 & 2 & 6 \\ -3 & -13 & 7 \end{bmatrix}$$

MULTIPLICATION OF A MATRIX BY A SCALAR

From the definition $C = A + B$ in which $c_{ij} = a_{ij} + b_{ij}$, it is natural to write for the case $A = B$ that $C = A + B = A + A = 2A$. It follows that $c_{ij} = a_{ij} + a_{ij} = 2a_{ij}$. Thus to multiply a matrix by a scalar, every element of the matrix is multiplied by the scalar.

MULTIPLICATION OF MATRICES.

Matrix multiplication is defined only under very special restrictions. The product $AB = C$ is defined only if the number of columns of A are equal to the number of rows of B. Thus if A is a matrix of order p by n and B is of order n by q, the multiplication is permitted and the product matrix C will be of order p by q. The elements of the product are defined as

$$c_{ij} = \sum_{k=1}^{n} (a_{ik}b_{kj}) \qquad \begin{aligned} i &= 1 \text{ to } p \\ j &= 1 \text{ to } q \end{aligned}$$

Example

$$\begin{bmatrix} 1 & 2 & 3 & 4 \\ 5 & 6 & 7 & 8 \\ 9 & 10 & 11 & 12 \end{bmatrix} \begin{bmatrix} 3 & 7 \\ 1 & 8 \\ 0 & 4 \\ 2 & 5 \end{bmatrix} = \begin{bmatrix} 13 & 55 \\ 37 & 151 \\ 61 & 247 \end{bmatrix}$$

Here

$$c_{3-2} = \sum_{k=1}^{4} (a_{3k}b_{k2}) = 9 \times 7 + 10 \times 8 + 11 \times 4 + 12 \times 5 = 247$$

Note that the multiplication of a matrix of order 3 by 4 by a matrix of order 4 by 2 results in a product matrix of order 3 by 2.

The multiplication of these two matrices in the reverse order is not permitted, since the condition that the number of columns of the first matrix must equal the number of rows of the second matrix is not satisfied.

Multiplication is not commutative in general, for even in the cases where multiplication is defined in the reverse order, it is generally not true that $AB = BA$. Consider, for example,

$$\begin{bmatrix} 1 & 2 & 3 \\ 4 & 5 & 6 \\ 7 & 8 & 9 \end{bmatrix} \begin{bmatrix} 0 & 5 & 7 \\ 2 & 1 & 4 \\ 3 & 3 & 2 \end{bmatrix} = \begin{bmatrix} 13 & 16 & 21 \\ 28 & 43 & 60 \\ 43 & 70 & 99 \end{bmatrix}$$

while

$$\begin{bmatrix} 0 & 5 & 7 \\ 2 & 1 & 4 \\ 3 & 3 & 2 \end{bmatrix} \begin{bmatrix} 1 & 2 & 3 \\ 4 & 5 & 6 \\ 7 & 8 & 9 \end{bmatrix} = \begin{bmatrix} 69 & 81 & 93 \\ 34 & 41 & 48 \\ 29 & 37 & 45 \end{bmatrix}$$

It is quite common in mathematical literature to speak of a matrix product by saying that AB is the result of premultiplying B by A, or that, A is postmultiplied by B. Multiplication is distributive. That is $C(A + B)D = CAD + CBD$. The associative law also applies $(AB)C = A(BC)$. The sequence of matrices in the product must be maintained.

THE IDENTITY MATRIX

The identity matrix is a square matrix with all elements of the major diagonal equal to unity and all other elements equal to zero. Pre- and postmultiplication of a square matrix A by the identity (or unit matrix) I results in a product matrix A which is the original matrix.

$$AI = IA = A$$

Here the unit matrix is the counterpart of unity in the multiplication of real numbers. The identity matrix (or unit matrix) is sometimes designated by I in England and the United States and by E in Continental Europe. In electrical engineering I and E have special significance; consequently U is sometimes used in this discipline to designate the unity matrix.

THE DETERMINANT OF A MATRIX

A determinant with the same array of elements as the elements of a square matrix is said to be the determinant of the matrix. A matrix also has associated with it minor determinants and cofactors.

It must be noted that both matrices and determinants are arrays but here the similarity ends. A determinant is always square and has a value which can be determined by one of several mathematical procedures [2–4]. A matrix is a rectangular array (not necessarily square) that does not have a value associated with it nor is a mathematical procedure involved. There is a complete and separate algebra for the treatment of matrices and determinants.

MINORS AND COFACTORS

A first minor for a determinant is the determinant that is formed when a particular row and column have been erased from the determinant. Thus the minor A_{ij} is the determinant composed of the matrix elements after deleting the ith row and the jth column. If

$$A = \begin{bmatrix} a_{11} & a_{12} & a_{13} & a_{14} \\ a_{21} & a_{22} & a_{23} & a_{24} \\ a_{31} & a_{32} & a_{33} & a_{34} \\ a_{41} & a_{42} & a_{43} & a_{44} \end{bmatrix}$$

then

$$A_{23} = \begin{vmatrix} a_{11} & a_{12} & a_{14} \\ a_{31} & a_{32} & a_{34} \\ a_{41} & a_{42} & a_{44} \end{vmatrix}$$

is the minor that results when the second row and third column are deleted from the original matrix.

A cofactor of a matrix is a minor multiplied by $(-1)^{i+j}$ which results in a checker board pattern of $+$ and $-$ signs beginning with $+$ in the upper left-hand corner.

The matrix formed by transposing rows and columns of cofactors is defined to be the adjoint of a matrix. Thus if the matrix $A = a_{ij}$ of order m by m is given, the matrix $B = A_{ji}(-1)^{j+i}$ where i and $j = 1, \ldots m$ is said to be the adjoint of A. Here A_{ji} is a minor.

MATRIX INVERSION

In matrix algebra the operation of division is not defined. Matrix inversion is the counterpart of division in the algebra of numbers. The inverse of a matrix is defined to be the matrix which when multiplied by the original matrix results in the unity matrix. This multiplication is commutative. Writing A^{-1} for the inverse of matrix A, the product $A^{-1}A = AA^{-1} = I$. For example,

$$\begin{bmatrix} 4 & 6 \\ 1 & 2 \end{bmatrix} \begin{bmatrix} 1 & -3 \\ -0.5 & 2 \end{bmatrix} = \begin{bmatrix} 1 & 0 \\ 0 & 1 \end{bmatrix}$$

If the matrix

$$\begin{bmatrix} 4 & 6 \\ 1 & 2 \end{bmatrix}$$

is the original matrix, then

$$\begin{bmatrix} 1 & -3 \\ -0.5 & 2 \end{bmatrix}$$

is said to be the inverse matrix.

In network analysis and in other studies it is very important to be able to obtain the inverse of a matrix. The classical method should be known by the student although it is seldom used in practice.

CLASSICAL MATRIX INVERSION

The inversion of a matrix

$$A = \begin{bmatrix} a_{11} & a_{12} & a_{13} \\ a_{21} & a_{22} & a_{23} \\ a_{31} & a_{32} & a_{33} \end{bmatrix}$$

by the classical method illustrates the procedure.

1. Obtain the transpose of the matrix

$$A_t = \begin{bmatrix} a_{11} & a_{21} & a_{31} \\ a_{12} & a_{22} & a_{32} \\ a_{13} & a_{23} & a_{33} \end{bmatrix}$$

2. Replace each element of the transpose by its cofactor.

$$A \text{ adjoint} = \begin{vmatrix} \begin{vmatrix} a_{22} & a_{32} \\ a_{23} & a_{33} \end{vmatrix} & -\begin{vmatrix} a_{12} & a_{32} \\ a_{13} & a_{33} \end{vmatrix} & \begin{vmatrix} a_{12} & a_{22} \\ a_{13} & a_{23} \end{vmatrix} \\ -\begin{vmatrix} a_{21} & a_{31} \\ a_{23} & a_{33} \end{vmatrix} & \begin{vmatrix} a_{11} & a_{31} \\ a_{13} & a_{33} \end{vmatrix} & -\begin{vmatrix} a_{11} & a_{21} \\ a_{13} & a_{23} \end{vmatrix} \\ \begin{vmatrix} a_{21} & a_{31} \\ a_{22} & a_{32} \end{vmatrix} & -\begin{vmatrix} a_{11} & a_{31} \\ a_{12} & a_{32} \end{vmatrix} & \begin{vmatrix} a_{11} & a_{21} \\ a_{12} & a_{22} \end{vmatrix} \end{vmatrix}$$

3. The elements of the adjoint matrix are divided by the value of the determinant of the original matrix. The result is the inverse of the original matrix.

The determinant of the matrix can be evaluated by several methods. Expansion by minors by the method of Gauss is used here.

$$D = \begin{vmatrix} a_{11} & a_{12} & a_{13} \\ a_{21} & a_{22} & a_{23} \\ a_{31} & a_{32} & a_{33} \end{vmatrix} = a_{11}\begin{vmatrix} a_{22} & a_{23} \\ a_{32} & a_{33} \end{vmatrix} - a_{12}\begin{vmatrix} a_{21} & a_{23} \\ a_{31} & a_{33} \end{vmatrix} + a_{13}\begin{vmatrix} a_{21} & a_{22} \\ a_{31} & a_{32} \end{vmatrix}$$

$$D = a_{11}(a_{22}a_{33} - a_{23}a_{32}) - a_{12}(a_{21}a_{33} - a_{23}a_{31}) + a_{13}(a_{21}a_{32} - a_{22}a_{31})$$

Then

$$A^{-1} = \begin{bmatrix} \dfrac{\begin{vmatrix} a_{22} & a_{32} \\ a_{23} & a_{33} \end{vmatrix}}{D} & -\dfrac{\begin{vmatrix} a_{12} & a_{32} \\ a_{13} & a_{33} \end{vmatrix}}{D} & \dfrac{\begin{vmatrix} a_{12} & a_{22} \\ a_{13} & a_{23} \end{vmatrix}}{D} \\ -\dfrac{\begin{vmatrix} a_{21} & a_{31} \\ a_{23} & a_{33} \end{vmatrix}}{D} & \dfrac{\begin{vmatrix} a_{11} & a_{31} \\ a_{13} & a_{33} \end{vmatrix}}{D} & -\dfrac{\begin{vmatrix} a_{11} & a_{21} \\ a_{13} & a_{23} \end{vmatrix}}{D} \\ \dfrac{\begin{vmatrix} a_{21} & a_{31} \\ a_{22} & a_{32} \end{vmatrix}}{D} & -\dfrac{\begin{vmatrix} a_{11} & a_{31} \\ a_{12} & a_{32} \end{vmatrix}}{D} & \dfrac{\begin{vmatrix} a_{11} & a_{21} \\ a_{12} & a_{22} \end{vmatrix}}{D} \end{bmatrix}$$

is the inverse of the given matrix A.

In selecting a method to be used for matrix inversion three factors must be taken into account:

1. The computational effort required, that is, the number of arithmetic operations in the inversion process.

2. Simplicity of the pattern of operations or the adaptability of the process to a computer program.

3. Auxiliary memory required during the calculation.

For matrices that can be held in core memory the best method that the author knows for satisfying these requirements is described by Shipley [5]. This technique requires no additional working memory besides that required for storage of the matrix. The matrix is inverted in place, and at the completion of the process the negative inverse matrix has replaced the original matrix.

SHIPLEY'S INVERSION

In this inversion method an operation that will be referred to as pivoting is performed on each major diagonal element in any sequence, using the operations to be described. At the time of pivoting the diagonal element used as the pivot can not be zero. Zero diagonal elements must be bypassed and not used as pivots until they have been modified to nonzero values by the process.

The operations in the pivoting process are performed once for each diagonal element as follows:

1. All elements not in the same row and column as the pivot element a_{kk} are modified by use of equation 2.2.

$$a'_{ij} = a_{ij} - a_{ik}\left(\frac{1}{a_{kk}}\right)a_{kj} \qquad i \neq k, \; j \neq k \tag{2.2}$$

2. Elements in the pivot row k are replaced by application of equation 2.3.

$$a'_{kj} = -\frac{a_{kj}}{a_{kk}} \qquad j \neq k \tag{2.3}$$

3. Elements in the pivot column k are replaced by application of equation 2.4.

$$a'_{ik} = -\frac{a_{ik}}{a_{kk}} \qquad i \neq k \tag{2.4}$$

4. The pivot element is replaced by

$$a'_{kk} = -\frac{1}{a_{kk}} \tag{2.5}$$

The process is repeated for each diagonal element taken in any sequence. When the process has been completed the negative inverse has replaced the original matrix.

In a single pivoting operation the following number of mathematical operations are involved.

Step	Equation	Additions	Multiplications	Divisions
1	2.2	$(n-1)^2$	$(n-1)^2$	$(n-1)^2$
2	2.3	—	—	$n-1$
3	2.4	—	—	$n-1$
4	2.5	—	—	1
Total		$(n-1)^2$	$(n-1)^2$	n^2

If the sequence of operations is rearranged, the number of calculations is reduced.

Step	Equation	Additions	Multiplications	Divisions
1	2.5	—	—	1
2	2.4	—	—	$n-1$
3	2.2	$(n-1)^2$	$(n-1)^2$	—*
4	2.3	—	—	$n-1$
Total		$(n-1)^2$	$(n-1)^2$	$2n-1$

In performing an inversion it is therefore necessary to carry out the pivoting calculation in the modified sequence of operations to achieve maximum computational efficiency.

* The modified column eliminates the division in equation 2.2.

Example of a Shipley Inversion   Find the inverse of

$$\text{matrix } A = \begin{bmatrix} 7 & 4 & 2 \\ 5 & 3 & 1 \\ 3 & 2 & 2 \end{bmatrix}$$

All major diagonals elements are nonzero. Pivoting may begin using any of the diagonal elements. Arbitrarily choosing a_{33}, the first pivoting is performed:

1. In the interest of minimizing the computational effort the diagonal element used as the pivot is modified first by equation 2.5.

$$a'_{33} = -\frac{1}{a_{33}} = -\frac{1}{2}$$

2. The elements of the pivoting column are modified by application of equation 2.4.

$$a'_{13} = -\frac{a_{13}}{a_{33}} = -\frac{2}{2} = -1$$

$$a'_{23} = -\frac{a_{23}}{a_{33}} = -\frac{1}{2}$$

3. The elements not in the pivot row and column are modified by using equation 2.2 and the precalculated values from step 2.

$$a'_{ij} = a_{ij} - a_{ik}\left(\frac{1}{a_{kk}}\right)a_{kj} = a_{ij} + a'_{ik}a_{kj}$$

$$a'_{11} = a_{11} - \left(\frac{a_{13}}{a_{33}}\right)(a_{31}) = 7 - (1)(3) = 4$$

$$a'_{12} = a_{12} - \left(\frac{a_{13}}{a_{33}}\right)(a_{32}) = 4 - (1)(2) = 2$$

$$a'_{21} = a_{21} - \left(\frac{a_{23}}{a_{33}}\right)(a_{31}) = 5 - \left(\frac{1}{2}\right)(3) = \frac{7}{2}$$

$$a'_{22} = a_{22} - \left(\frac{a_{23}}{a_{33}}\right)(a_{32}) = 3 - \left(\frac{1}{2}\right)(2) = 2$$

4. The elements in the pivot row are modified by use of equation 2.3.

$$a'_{31} = -a_{31}\left(\frac{1}{a_{33}}\right) = -\frac{3}{2}$$

$$a'_{32} = -a_{32}\left(\frac{1}{a_{33}}\right) = -1$$

After the first pivoting operation the matrix is

$$A_1 = \begin{bmatrix} 4 & 2 & -1 \\ \frac{7}{2} & 2 & -\frac{1}{2} \\ -\frac{3}{2} & -1 & -\frac{1}{2} \end{bmatrix}$$

Since a_{11} and a_{22} are both nonzero and have not been used as a pivot, either can now be used. Arbitrarily a_{22} is chosen as the pivot element.

$$a'_{22} = -\tfrac{1}{2}$$

The pivot column becomes

$$a'_{12} = -\frac{a_{12}}{a_{22}} = -1$$

$$a'_{32} = -\frac{a_{32}}{a_{22}} = -\left(\frac{-1}{2}\right) = \frac{1}{2}$$

The elements not in the pivot row and column are

$$a'_{11} = a_{11} - \left(\frac{a_{12}}{a_{22}}\right)(a_{21}) = 4 - (1)\left(\frac{7}{2}\right) = \frac{1}{2}$$

$$a'_{13} = a_{13} - \left(\frac{a_{12}}{a_{22}}\right)(a_{23}) = -1 - (1)\left(-\frac{1}{2}\right) = -\frac{1}{2}$$

$$a'_{31} = a_{31} - \left(\frac{a_{32}}{a_{22}}\right)(a_{21}) = -\frac{3}{2} + \left(\frac{1}{2}\right)\left(\frac{7}{2}\right) = \frac{1}{4}$$

$$a'_{33} = a_{33} - \left(\frac{a_{32}}{a_{22}}\right)(a_{23}) = -\frac{1}{2} + \left(\frac{1}{2}\right)\left(-\frac{1}{2}\right) = -\frac{3}{4}$$

Modification of the pivot row gives

$$a'_{21} = -\frac{a_{21}}{a_{22}} = -\frac{7/2}{2} = -\frac{7}{4}$$

$$a'_{23} = -\frac{a_{23}}{a_{22}} = \frac{1/2}{2} = \frac{1}{4}$$

The matrix after the second pivoting operation is

$$A_2 = \begin{bmatrix} \frac{1}{2} & -1 & -\frac{1}{2} \\ -\frac{7}{4} & -\frac{1}{2} & \frac{1}{4} \\ \frac{1}{4} & \frac{1}{2} & -\frac{3}{4} \end{bmatrix}$$

Pivoting on the remaining diagonal term that has not been used as a pivot gives the negative inverse matrix. The calculations in the third pivoting are left as an exercise for the student.

$$A_3 = -A^{-1} = \begin{bmatrix} -2 & 2 & 1 \\ \frac{7}{2} & -4 & -\frac{3}{2} \\ -\frac{1}{2} & +1 & -\frac{1}{2} \end{bmatrix}$$

$$A^{-1} = \begin{bmatrix} 2 & -2 & -1 \\ -\frac{7}{2} & 4 & \frac{3}{2} \\ \frac{1}{2} & -1 & \frac{1}{2} \end{bmatrix}$$

Verification that $AA^{-1} = A^{-1}A = I$ is also left as an exercise for the student.

Justification of the Shipley Inversion Consider the set of linear equations

$$a_{11}X_1 + a_{12}X_2 + a_{13}X_3 = b_1$$

$$a_{21}X_1 + a_{22}X_2 + a_{23}X_3 = b_2 \qquad (2.6)$$

$$a_{31}X_1 + a_{32}X_2 + a_{33}X_3 = b_3$$

These equations can be written in matrix form as

$$
\begin{bmatrix} a_{11} & a_{12} & a_{13} \\ a_{21} & a_{22} & a_{23} \\ a_{31} & a_{32} & a_{33} \end{bmatrix} \begin{bmatrix} X_1 \\ X_2 \\ X_3 \end{bmatrix} = \begin{bmatrix} b_1 \\ b_2 \\ b_3 \end{bmatrix} \tag{2.7}
$$

or more simply as

$$ AX = b \tag{2.8} $$

The X's are the independent variables and the b's are the dependant variables. If an inverse A^{-1} can be found for the matrix A the equation 2.8 can be multiplied through on the left by this inverse and would yield $A^{-1}AX = IX = X = A^{-1}b$. By inversion the independent and dependant variables have changed roles.

Dividing the first equation of (2.6) through by a_{11} and solving for X_1 explicitly in terms of the other variables and coefficients gives

$$ \frac{b_1}{a_{11}} - \frac{a_{12}}{a_{11}} X_2 - \frac{a_{13}}{a_{11}} X_3 = X_1 $$

Substituting this value of X_1 in the second and third equations of (2.6) and rewriting all three equations again gives

$$ \frac{b_1}{a_{11}} - \frac{a_{12}}{a_{11}} X_2 - \frac{a_{13}}{a_{11}} X_3 = X_1 $$

$$ a_{21}\left(\frac{b_1}{a_{11}} - \frac{a_{12}}{a_{11}} X_2 - \frac{a_{13}}{a_{11}} X_3 \right) + a_{22}X_2 + a_{23}X_3 = b_2 \tag{2.9} $$

$$ a_{31}\left(\frac{b_1}{a_{11}} - \frac{a_{12}}{a_{11}} X_2 - \frac{a_{13}}{a_{11}} X_3 \right) + a_{32}X_2 + a_{33}X_3 = b_3 $$

Collecting terms gives

$$ \frac{1}{a_{11}}b_1 - \frac{a_{12}}{a_{11}} X_2 - \frac{a_{13}}{a_{11}} X_3 = X_1 $$

$$ \frac{a_{21}}{a_{11}}b_1 + \left(a_{22} - \frac{a_{21}a_{12}}{a_{11}} \right)X_2 + \left(a_{23} - \frac{a_{21}a_{13}}{a_{11}} \right)X_3 = b_2 \tag{2.10} $$

$$ \frac{a_{31}}{a_{11}}b_1 + \left(a_{32} - \frac{a_{31}a_{12}}{a_{11}} \right)X_2 + \left(a_{33} - \frac{a_{31}a_{13}}{a_{11}} \right)X_3 = b_3 $$

This can be written in matrix form as

$$
\begin{bmatrix}
\dfrac{1}{a_{11}} & -\dfrac{a_{12}}{a_{11}} & -\dfrac{a_{13}}{a_{11}} \\[2mm]
\dfrac{a_{21}}{a_{11}} & \left(a_{22}-\dfrac{a_{21}a_{12}}{a_{11}}\right) & \left(a_{23}-\dfrac{a_{21}a_{13}}{a_{11}}\right) \\[2mm]
\dfrac{a_{31}}{a_{11}} & \left(a_{32}-\dfrac{a_{31}a_{12}}{a_{11}}\right) & \left(a_{33}-\dfrac{a_{31}a_{13}}{a_{11}}\right)
\end{bmatrix}
\begin{bmatrix} b_1 \\ X_2 \\ X_3 \end{bmatrix}
=
\begin{bmatrix} X_1 \\ b_2 \\ b_3 \end{bmatrix}
$$

Multiplying the elements in the first column of the coefficient matrix by -1 and the first element in the column vector of independent variables also by -1 gives

$$
\begin{bmatrix}
-\dfrac{1}{a_{11}} & -\dfrac{a_{12}}{a_{11}} & -\dfrac{a_{13}}{a_{11}} \\[2mm]
-\dfrac{a_{21}}{a_{11}} & \left(a_{22}-\dfrac{a_{21}a_{12}}{a_{11}}\right) & \left(a_{23}-\dfrac{a_{21}a_{13}}{a_{11}}\right) \\[2mm]
-\dfrac{a_{31}}{a_{11}} & \left(a_{32}-\dfrac{a_{31}a_{12}}{a_{11}}\right) & \left(a_{33}-\dfrac{a_{31}a_{13}}{a_{11}}\right)
\end{bmatrix}
\begin{bmatrix} -b_1 \\ X_2 \\ X_3 \end{bmatrix}
=
\begin{bmatrix} X_1 \\ b_2 \\ b_3 \end{bmatrix}
$$

$$(2.11)$$

Equation 2.11 corresponds exactly to what is obtained when pivoting is performed in Shipley's method using a_{11} as the pivot. It is seen that the first independent variable has changed places with the negative of the first dependant variable.

Pivoting on each diagonal element in turn will exchange all of the dependant and independent variables with a change of sign. The result can be written as

$$
\begin{bmatrix}
c_{11} & c_{12} & c_{13} \\
c_{21} & c_{22} & c_{23} \\
c_{31} & c_{32} & c_{33}
\end{bmatrix}
\begin{bmatrix} -b_1 \\ -b_2 \\ -b_3 \end{bmatrix}
=
\begin{bmatrix} X_1 \\ X_2 \\ X_3 \end{bmatrix}
\qquad \text{or} \qquad [C][-b]=[X]
$$

The matrix

$$C = \begin{bmatrix} c_{11} & c_{12} & c_{13} \\ c_{21} & c_{22} & c_{23} \\ c_{31} & c_{32} & c_{33} \end{bmatrix}$$

is the negative of the inverse of A since $A^{-1}b = X$.

PARTITIONING OF MATRICES

Economy in the computational effort may often result by resorting to the technique known as partitioning. This is specially true if the order of the matrix is very large. It can happen that the matrix being used in the computation must be partitioned because the complete matrix could not be held in the high-speed memory. A matrix can be partitioned in any convenient manner that is suited to the calculation being made. For example,

$$\begin{bmatrix} a_{11} & a_{12} & a_{13} & a_{14} \\ a_{21} & a_{22} & a_{23} & a_{24} \\ a_{31} & a_{32} & a_{33} & a_{34} \\ a_{41} & a_{42} & a_{43} & a_{44} \end{bmatrix} = \left[\begin{array}{cc|cc} a_{11} & a_{12} & a_{13} & a_{14} \\ a_{21} & a_{22} & a_{23} & a_{24} \\ \hline a_{31} & a_{32} & a_{33} & a_{34} \\ a_{41} & a_{42} & a_{43} & a_{44} \end{array} \right]$$

$$= \left[\begin{array}{c|ccc} a_{11} & a_{12} & a_{13} & a_{14} \\ \hline a_{21} & a_{22} & a_{23} & a_{24} \\ a_{31} & a_{32} & a_{33} & a_{34} \\ a_{41} & a_{42} & a_{43} & a_{44} \end{array} \right]$$

In the work which follows partitioning is restricted to the case where all diagonal submatrices are square but not necessarily of the same order. This is a special case of partitioning known as quasi-partitioning. The matrix

2.12 is an example of quasi-partitioning:

$$
\left[
\begin{array}{cccccc}
a_{11} & a_{12} & 0 & 0 & 0 & 0 \\
a_{21} & a_{22} & 0 & 0 & 0 & 0 \\
0 & 0 & b_{11} & b_{12} & b_{13} & 0 \\
0 & 0 & b_{21} & b_{22} & b_{23} & 0 \\
0 & 0 & b_{31} & b_{32} & b_{33} & 0 \\
0 & 0 & 0 & 0 & 0 & c_{11}
\end{array}
\right] \tag{2.12}
$$

Exercises

1. Invert the matrix

$$
A = \left[
\begin{array}{ccc}
7 & 4 & 2 \\
5 & 3 & 1 \\
3 & 2 & 2
\end{array}
\right]
$$

by the classical method. How many multiplications were necessary?

2. How many multiplications does it require to invert the matrix in Exercise 1 by the Shipley method discussed in the text?

References

1. **R. A. Frazer, W. J. Duncan, and A. R. Collar,** *Elementary Matrices*, Cambridge Press, 1955.
2. **V. N. Faddeeva,** *Computational Methods of Linear Algebra*, Dover, 1959.
3. **O. Pedoe,** *A Geometric Introduction to Linear Algebra*, Wiley, 1963.
4. **L. Dickson,** *Linear Transformations and Matrices*, Prentice-Hall, 1967.
5. **R. B. Shipley and D. Coleman,** A new direct matrix inversion method, *Trans. AIEE C & E*, Vol. 78, (November 1959), pp. 568–572.

3

Three-Phase
Short-Circuit Calculations

An electrical network under short-circuit conditions can be considered a network supplied by several sources (generators) with a single load connected to the system at the node subjected to the short circuit. The normal customer load currents are usually ignored, since they are small compared to the short-circuit current. Generally this simplification does not impare the accuracy of the short-circuit study. This is equivalent to the structural analysis of a bridge supported by several piers and subjected to a single concentrated load, with the weight of the individual members of the structure being ignored. The remainder of this text concentrates on the analysis of electrical network problems, but one should remember that the techniques developed here apply equally well to structures.

For a treatment of the reverse approach see Ref. 6. In the introduction to Chapter 9 these authors write, "The language of the dynamics of material systems will be used throughout, but the treatment can, for instance, be applied equally well to electrical systems." The complete short-circuit analysis of a network is possible by simple arithmetic operations, as soon as all the node voltages have been determined for the particular fault condition. Coombe and Lewis solved the node voltages in short-circuit studies by the Gauss-Seidel iterative approach developed by Ward and Hale for the solution of normal power flow problems [1,2]. The method is poorly suited to short-circuit studies, since each fault condition requires an iterative solution. A complete analysis of a 1000-node (bus) system may require as many as 20,000 to 30,000 fault conditions, each of which would be an entirely new iterative procedure. The short-circuit analysis of very large electrical systems is achieved most efficiently by use of the Z-bus matrix [3].

DESCRIPTION OF THE Z-BUS MATRIX

The Z-bus matrix contains the driving point impedance of every node with respect to a reference node that has been chosen arbitrarily. The driving point impedance of a node is the equivalent impedance between it and the reference. The Z-bus matrix also contains the transfer impedance between each bus of the system and every other bus with respect to the reference bus. The transfer impedances are determined by computing the voltages that exist on each of the other buses of the system, with respect to the reference, when a particular bus of the system is driven by an injection current of unity (see Fig. 3.1).

The matrix equation relating the Z-bus matrix, the currents injected into the nodes, and the node voltages is

$$ZI = E \tag{3.1}$$

It was recognized very early that if the Z-bus matrix with the reference bus chosen as the common bus behind the generator transient reactances was available, the complete short-circuit analysis of the network could be readily obtained with a small amount of additional computation. Remembering, as it was indicated earlier, that a network under fault conditions could be considered to have a single node current, one can write the matrix equation as

$$
\begin{bmatrix}
Z_{11} & Z_{12} & Z_{13} & \cdots & Z_{1k} & \cdots & Z_{1n} \\
Z_{21} & Z_{22} & Z_{23} & \cdots & Z_{2k} & \cdots & Z_{2n} \\
\cdots & \cdots & \cdots & \cdots & \cdots & \cdots & \cdots \\
\cdots & \cdots & \cdots & \cdots & \cdots & \cdots & \cdots \\
Z_{k1} & Z_{k2} & Z_{k3} & \cdots & Z_{kk} & \cdots & Z_{kn} \\
\cdots & \cdots & \cdots & \cdots & \cdots & \cdots & \cdots \\
Z_{n1} & Z_{n2} & Z_{n3} & \cdots & Z_{nk} & \cdots & Z_{nn}
\end{bmatrix}
\begin{bmatrix}
0 \\
0 \\
\vdots \\
I_k \\
\vdots \\
0
\end{bmatrix}
=
\begin{bmatrix}
E_1 \\
E_2 \\
\vdots \\
E_k \\
\vdots \\
E_n
\end{bmatrix}
\tag{3.2}
$$

where the network is subjected to the single current injection I_k into node k, which is the bus that is in fault condition. Obviously, column k enables the voltage profile to be determined for the network, when a short-circuit occurs on node k, provided that I_k has been, or can be, determined. The element Z_{kk} is the driving point impedance of bus k. The off diagonal

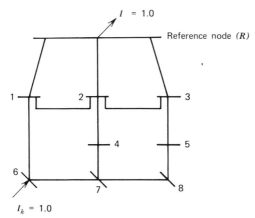

Fig. 3.1. Driving point impedance: $E_{k-r} = IZ_{kk}$; $I_k = 1.0$; $Z_{kk} = E_{k-r}$. Transfer impedance: $I_k = 1.0$; $E_{ir} = IZ_{ik}$; $Z_{ik} = E_{ir}$.

elements Z_{ik} are the transfer impedance between the other buses and bus k.

In short-circuit calculations it is customary to assume that all generators connected to the network are operating with 1.0 per unit voltage behind their internal reactances. This common point behind the generator reactances is used as reference. The network can therefore be considered to be supplied by a single common source (see Fig. 3.2).*

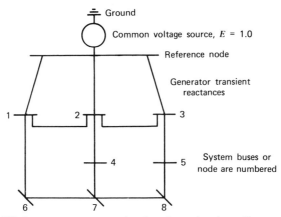

Fig. 3.2. Simplified system representation for short circuit studies.

*The analogous situation occurs in the analysis of the network of a steel bridge. The reaction at each pier ultimately derives its capability from a single source, the earth.

The use of 1.0 pu voltage behind internal machine reactance can be justified by applying the Helmholtz-Thevenin theorem. The prefault open circuit voltage at the point of fault is approximately 1.0 pu and may therefore be assumed to be that value. The short-circuit impedance of the network from the point of fault is determined by the impedance of the network elements (including the internal machine impedances) with the internal source voltages all short-circuited. The fault current is calculated as the superposition of two sources: the current due to the internal source voltages, which are zero; and the current due to a superposed voltage source which will reduce the fault point voltage to zero. This voltage is obviously equal to the negative of the prefault voltage and is thus equal to -1.0 pu from ground to the point of fault. But this gives the same current as $+1.0$ pu voltage from ground to a common bus behind transient reactances.

When any node is short-circuited, it is connected to ground. Full voltage is therefore applied between the reference node and the node subjected to the fault condition. For example, for a fault on node 6, the diagram can be drawn as shown in Fig. 3.3.

Since the matrix elements of the Z-matrix of equation 3.2 are the driving point impedances (diagonal elements) and transfer impedances (off-diagonal elements) with respect to the reference bus, the voltages of equation 3.2 will all be measured with respect to the reference node behind the generator transient reactances. The reference node is therefore at zero potential with respect to itself, but is at full voltage with respect to ground in the actual network. The voltage obtained from equation 3.2 for the bus

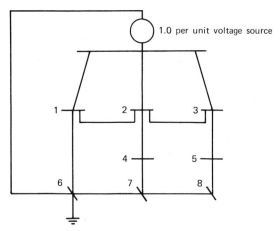

Fig. 3.3.   Node 6 is in short circuit condition.

under fault condition is the full voltage of the generators with respect to the reference. In the actual system the bus that is short-circuited is at zero potential with respect to ground. This difference in voltage, depending on the point of reference, should cause no difficulty but must be taken into consideration in expressing the results of a calculation.

$$E_p^g = 1 - E_p^m$$

where E_p^g is the voltage of bus p with ground as the reference, as it would be measured in the actual system. The E_p^m is the voltage obtained from the matrix calculation of equation 3.2 and is measured with respect to the reference bus behind the generator transient reactances.

When bus k is in short-circuit condition, the constraint that full voltage is applied to bus k will be satisfied by injecting a current I_k that is determined by equation 3.3.

$$I_k = \frac{1}{Z_{kk}} \qquad (3.3)$$

See also equation 3.2.

The total fault current for any bus is therefore obtained by taking the reciprocal of the corresponding diagonal element of the Z-matrix. The voltages that appear on the other buses of the system, when bus k is in fault condition, depend on the transfer impedances as given by the off diagonal elements of column k of the Z-bus matrix. For example, the voltage with respect to the reference on bus p for a short circuit on bus k would be given by equation 3.4.

$$E_p = Z_{pk} \frac{1}{Z_{kk}} \qquad (3.4)$$

The current flow from bus p to bus q over the line $p - q$ whose impedance is $Z_{\text{line}pq}$ is

$$I_{pq} = \frac{Z_{qk} - Z_{pk}}{Z_{\text{line}pq}} \frac{1}{Z_{kk}} \qquad (3.5)$$

The total fault current for a fault on any bus is obtained by equation 3.3, and the flow in any line for a short circuit on a particular bus is obtained by equation 3.5. The complete analysis of the system is obtained by these simple arithmetic operations once the Z-matrix has been obtained.

THE Z-MATRIX BUILDING ALGORITHM

To compute the driving point and transfer impedance matrix of a completely assembled transmission system would be utterly impossible. However, it is possible, by rather simple means, to modify the Z-matrix of a system for the addition of a single line. In this way the system can be assembled by starting with a system of a single transmission line, adding one line at a time, modifying the matrix for each line addition, and assembling the desired system and the matrix that corresponds to the system [3].

DATA PREPARATION

A system diagram is drawn. The junction points, where two or more transmission lines, transformers, or generator impedances are connected, are assigned a unique bus (node) number. The number zero is reserved for the reference bus. In short-circuit studies the reference bus is selected as the common point behind all generator reactances. (In other studies the reference bus may be selected as ground or a bus of the system. See Chapter 6.)

Data are prepared by describing each element of the transmission system by the two buses at the ends of the line and its impedance on a common per unit base. These data are sequenced by an algorithm from a random ordering to a sequence such that as each line is selected from the data list for processing, it can be connected to the system that has been assembled. The first line in the list must be one from the reference to some bus of the system to provide a path to the reference for current injected into any node of the network being assembled.

Each line selected from the list must fall into one of three categories.

1. A line from the reference to a new bus.
2. A radial line from an existing bus to a new bus.
3. A line between two buses already included in the system, (a loop closing line).

Three different routines are required to modify the matrix for the addition of a line to the system depending on the type of line to be added.

A REFERENCE LINE TO A NEW BUS

A line from the reference to a new bus of the system is identified by finding that one bus is the reference bus and the other bus is not in the system that has been assembled.

Current injected into the new bus k which is connected by a radial line to the reference, will produce no voltage on the other buses of the system (see Fig. 3.4).

Injection of current into any bus of the system that had been assembled will produce no voltage on the new bus k. All off diagonal elements of the new row and column are therefore zero.

$$Z_{ik} = Z_{ki} = 0 \qquad i \neq k \tag{3.6}$$

The driving point impedance of the new bus is the impedance of the new line being added. The diagonal element of a new matrix axis corresponding to bus k is given by equation 3.7.

$$Z_{kk} = Z_{\text{line } o-k} \tag{3.7}$$

For the addition of a radial line from the reference to a new bus, augment the matrix by a row and column of zeros. The diagonal element of this new axis is the impedance of the new line being added. The bus number k is added to the list of buses that comprise the system.

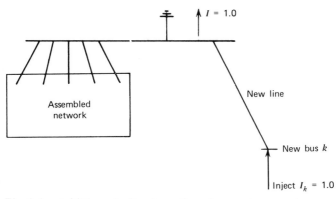

Fig. 3.4. Addition of a line from the reference to a new bus.

ADDITION OF A RADIAL LINE TO A NEW BUS

A radial line from a bus of the system to a new bus is identified by finding that neither bus describing the line is the reference and only one bus describing the line is included in the system. Injection of unit current into bus q produces voltages on all other buses of the system that are identical to the voltages that would be produced if current was injected into bus p (see Fig. 3.5).

$$
\begin{aligned}
Z_{qk} &= Z_{pk} \\
Z_{kq} &= Z_{kp}
\end{aligned} \qquad k \neq q \tag{3.8}
$$

The driving point impedance of bus q is equal to the driving point impedance of bus p plus the impedance of the line being added (see Fig. 3.5).

$$
Z_{qq} = Z_{pp} + Z_{\text{line } pq} \tag{3.9}
$$

A new axis is added to the matrix corresponding to the new bus q. The off-diagonal elements of the new row and column are the same as the elements of the row and column of bus p of the existing system. The diagonal element is obtained from equation 3.9. Bus q is added to the system bus list.

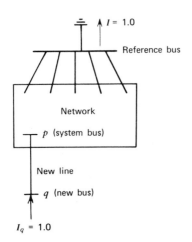

Fig. 3.5. Addition of a radial line from a system bus to a new bus.

ADDITION OF A LOOP CLOSING LINE

A loop closing line is identified by finding that both buses describing the line are included in the list of buses of the system that has been assembled. Addition of a loop closing line does not add a new node to the system that will provide a new axis for the matrix. There is, however, an additional constraint. The response of the system to the introduction of a current of unity into the loop created by the addition of this new line is a constraint that must be satisfied.

Injection of a current of unity into bus p causes voltages to appear at every bus of the system that are identical to the Z-matrix elements of column p of the matrix (see Fig. 3.6). Injection of a current into bus q of -1.0 produces voltages on the buses of the system equal to the Z-matrix elements of column q but of opposite sign. A loop current of unity can be considered to be a current of $I_p = 1.0$ and $I_q = -1.0$ acting in concert. The voltages appearing on the system buses are the difference of the columns corresponding to buses p and q as given in equation 3.10.

$$
\begin{bmatrix}
Z_{11} & Z_{12} \cdots Z_{1p} \cdots Z_{1q} \cdots Z_{1n} \\
\cdots \cdots \cdots Z_{2p} \cdots Z_{2q} \cdots Z_{2n} \\
\cdots \cdots \cdots Z_{3p} \cdots Z_{3q} \cdots \cdots \\
\cdots \cdots \cdots \cdots \cdots \cdots \cdots \\
\cdots \cdots \cdots Z_{np} \cdots Z_{nq} \cdots Z_{nn}
\end{bmatrix}
\begin{bmatrix}
0 \\
0 \\
I_p = 1.0 \\
I_q = -1.0 \\
0
\end{bmatrix}
=
\begin{bmatrix}
Z_{1p} - Z_{1q} \\
Z_{2p} - Z_{2q} \\
\vdots \\
Z_{np} - Z_{nq}
\end{bmatrix}
$$

$$(3.10)$$

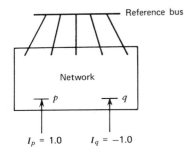

Fig. 3.6. Simulation of a loop closing line by two injection currents.

The amount of voltage that must be introduced into the loop to cause unit current to circulate in the loop created by addition of the new line can be computed by 3.11, as is evident in Fig. 3.7.

$$E_{\text{loop}} = I_{\text{loop}}(Z_{pp} - Z_{pq} + Z_{qq} - Z_{pq} + Z_{\text{line-}pq}) \qquad (3.11)$$

The driving point impedance of the loop, $Z_{\text{loop-loop}}$, is determined by setting $I_{\text{loop}} = 1.0$ in (3.11).

$$Z_{\text{loop-loop}} = Z_{pp} + Z_{qq} - 2Z_{pq} + Z_{\text{line-}pq} \qquad (3.12)$$

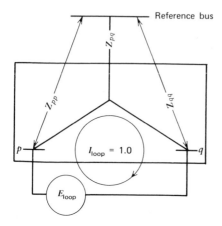

Fig. 3.7. Determination of the driving point impedance of the loop.

A loop axis is added to the Z-matrix of equation 3.10 in which

$$Z_{i\text{-loop}} = Z_{ip} - Z_{iq} \qquad i \neq \text{loop}$$

$$Z_{\text{loop}-i} = Z_{pi} - Z_{qi} \qquad i \neq \text{loop} \qquad (3.13)$$

The diagonal element is obtained from 3.12. The loop axis is eliminated from the matrix by a Kron reduction [4] by application of equation 3.15. All elements not in the loop row or column are modified. The loop row and column are erased. The list of system buses remains unchanged.

The Kron reduction is very simple if a single axis is being eliminated. This is shown later in the illustration of the building algorithm.

KRON REDUCTION OF A MATRIX

Consider the matrix of Fig. 3.8. The submatrix Z_1 can be modified to

$$\begin{bmatrix} Z_1 & Z_2 \\ \hline Z_3 & Z_4 \end{bmatrix}$$

Fig. 3.8. Partitioned matrix to be reduced.

reflect the changes in the network that take place when the axes corresponding to the rows of Z_3 and columns of Z_2 are eliminated. This modified matrix Z_1' can be viewed as a Kron network reduction or an algebraic elimination. Kron reduction is given by

$$Z_1' = Z_1 - Z_2 Z_4^{-1} Z_3 \tag{3.15}$$

The validity of the reduction 3.15 can be proved by considering the matrix equation 3.16

$$\begin{bmatrix} A_1 & A_2 \\ A_3 & A_4 \end{bmatrix} \begin{bmatrix} X_1 \\ X_2 \end{bmatrix} = \begin{bmatrix} B_1 \\ B_2 \end{bmatrix} \tag{3.16}$$

in which A_1, A_2, A_3, A_4 can be thought of as matrices or single coefficients and X_1, X_2, B_1 and B_2 are vectors or single variables, respectively.

Equation 3.16 in the expanded form is

$$A_1 X_1 + A_2 X_2 = B_1 \tag{3.17}$$

$$A_3 X_1 + A_4 X_2 = B_2 \tag{3.18}$$

Rewriting (3.18) gives

$$A_4 X_2 = B_2 - A_3 X_1 \tag{3.19}$$

Premultiplication by A_4^{-1} gives

$$X_2 = A_4^{-1} B_2 - A_4^{-1} A_3 X_1 \tag{3.20}$$

Substitution of X_2 into (3.17) gives

$$A_1 X_1 + A_2 (A_4^{-1} B_2 - A_4^{-1} A_3 X_1) = B_1 \tag{3.21}$$

Rearranging and collecting terms give

$$(A_1 - A_2 A_4^{-1} A_3) X_1 = B_1 - A_2 A_4^{-1} B_2 \tag{3.22}$$

The coefficient of the unknown X_1 is seen to be the result of a Kron reduction on A_1 as given in (3.15) when the unknown X_2 is eliminated.

AN ILLUSTRATIVE EXAMPLE OF THE BUILDING ALGORITHM

Consider the network shown in Fig. 3.9 and the following data.

System Data	
Line	X (pu)
0-1	0.010
0-2	0.015
1-2	0.084
0-3	0.005
2-3	0.122
2-4	0.084
3-5	0.037
1-6	0.126
6-7	0.168
4-7	0.084
5-8	0.037
7-8	0.140

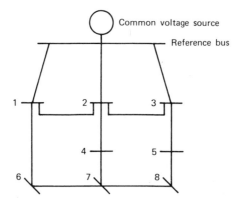

Fig. 3.9.

The matrix of the network is built by assembling the system one line at a time, and by modifying the matrix to reflect the change in the network equivalent impedances by the addition of the line. The line list has been reordered from a random list to a sequence such that it is possible to connect each line to the system when it is selected from the list for processing.

Addition of the First Line The first line must always be a line connected to the reference. Thus the line 0-1 can be the first line processed. At this point no network, no matrix, and no entries in the bus list describe the system. The buses 0 and 1 are examined and compared with the system bus list to determine the type of line and the algorithm to be used. The line is found to be a line connected to the reference and to bus 1. Bus 1 is compared with the list of buses in the system. At this point there are no buses in the network. The line is, therefore, a degenerate case of the addition of a line from the reference to a new bus (see Fig. 3.4).

It is impossible to add a row and column of zeroes to the matrix, since there is no matrix at this point. The diagonal element of the new axis is the impedance of the line being added. The new bus is added to the bus list. After adding this first line we have:

$$1$$
$$\text{matrix} = 1[0.01] \qquad \text{bus list} = 1$$

The system diagram is shown below.

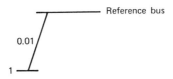

The matrix tells us that unit current injected into bus 1 would cause a voltage of 0.01 to appear on bus 1 when measured with respect to reference bus.

Addition of the Second Line The next line (0-2) is selected from the data list for processing. Examination of the bus numbers 0 and 2 and comparison with the system bus list show that this line is from the reference bus (0) to a new bus 2. This is also a type 1 line. Augment the matrix with a row and column of zeroes. The diagonal element of the new axis is the impedance of the new line. Add bus 2 to the bus list.

$$\text{matrix} = \begin{array}{c} \\ 1 \\ 2 \end{array} \begin{array}{cc} 1 & 2 \\ \begin{bmatrix} 0.01 & 0 \\ 0 & 0.015 \end{bmatrix} \end{array} \qquad \text{bus list} = 1, 2$$

The system diagram is shown below.

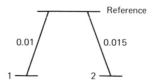

The matrix shows us that injection of unit current into bus 1 and out at the reference will cause a voltage of 0.01 to appear at bus 1 and a voltage of zero to appear on bus 2. Injection of unit current into bus 2 will produce a voltage of 0.015 on bus 2 and a voltage of zero on bus 1.

Addition of the Third Line The next line (1-2) is selected for processing. Examination of the bus numbers shows that this line is not a reference bus line. Comparison of the bus numbers of the line with the system bus numbers verify that the line is a type 3 (loop closing) line.

The matrix is augmented by a loop row and column by taking the difference of the rows (and columns) corresponding to buses 1 and 2, equation 3.13. The diagonal element is obtained by equation 3.12.

$$
\text{matrix} \quad
\begin{array}{c}
1 \\
2 \\
\text{loop}
\end{array}
\begin{array}{ccc}
1 & 2 & \text{loop} \\
\left[\begin{array}{ccc}
0.01 & 0 & 0.01 \\
0 & 0.015 & -0.015 \\
0.01 & -0.015 & 0.109
\end{array}\right]
\end{array}
\quad \text{bus list } 1,2
$$

The system diagram is shown below.

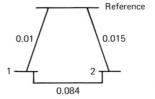

Fig. 3.10.

The matrix is reduced by application of the Kron reduction using equation 3.15. As was pointed out, the Kron reduction is greatly simplified when a single axis is involved in the reduction. In that case Z_4^{-1} becomes

$1/Z_4$ and a matrix inversion is not required. The modification of the elements not in the loop row and column can be most easily carried out element by element rather than by applying equation 3.15 as a matrix equation.

It can be easily verified that the modification of an element; Z_{ij} is simply:

$$Z'_{ij} = Z_{ij} - Z_{i\text{-loop}}\left(\frac{1}{Z_{\text{loop-loop}}}\right)Z_{\text{loop-}j} \qquad (3.23)$$

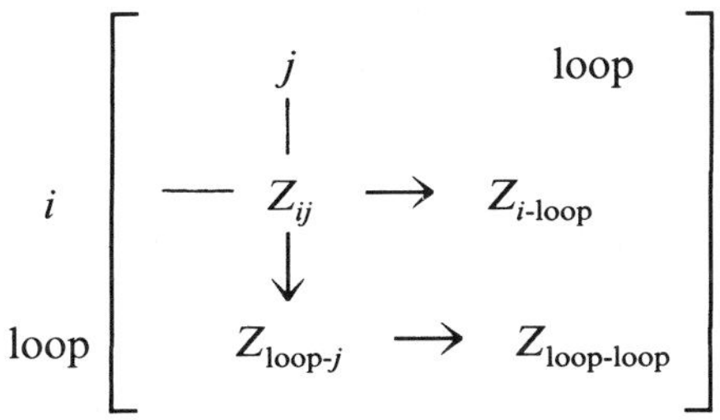

$$
\begin{bmatrix}
& j & & \text{loop} \\
& | & & \\
i & -\!\!- \; Z_{ij} & \rightarrow & Z_{i\text{-loop}} \\
& \downarrow & & \\
\text{loop} & Z_{\text{loop-}j} & \rightarrow & Z_{\text{loop-loop}}
\end{bmatrix}
$$

Fig. 3.11 Matrix elements used in modifying element Z_{ij} by the Kron reduction.

Application of equation 3.23 is used to modify all elements not in the loop axis of the matrix. The loop axis is then erased.

$$Z'_{11} = Z_{11} - Z_{1\text{-loop}}\left(\frac{1}{Z_{\text{loop-loop}}}\right)Z_{\text{loop-}1} = 0.01 - \frac{(0.01)(0.01)}{0.109}$$

$$= 0.01 - 0.00091743 = 0.00908257$$

$$Z'_{12} = Z_{12} - Z_{1\text{-loop}}\left(\frac{1}{Z_{\text{loop-loop}}}\right)Z_{\text{loop-}2} = 0 - \frac{(0.01)(-0.015)}{0.109}$$

$$= 0 + 0.00137615 = 0.00137615$$

$$Z'_{22} = 0.015 - \frac{(-0.015)(-0.015)}{0.109} = 0.015 - 0.00206421$$

$$= 0.01293579$$

The modified matrix is

$$
\text{matrix} \quad
\begin{array}{c}
 \\
1 \\
2
\end{array}
\begin{bmatrix}
1 & 2 \\
0.00908257 & 0.00137615 \\
0.00137615 & 0.01293579
\end{bmatrix}
\quad \text{bus list } 1, 2
$$

The Matrix Reduction—A Delta-Star Conversion The matrix reduction can be viewed as a delta-star reduction of the network. The network of Fig. 3.10 can be converted to the equivalent star by the standard delta-star conversion.

$$Z_1 = \frac{(0.01)(0.015)}{0.01 + 0.015 + 0.084} = \frac{(0.01)(0.015)}{0.109} = 0.00137615$$

$$Z_2 = \frac{(0.01)(0.084)}{0.109} = 0.00770642$$

$$Z_3 = \frac{(0.015)(0.084)}{0.109} = 0.01155964$$

If the network of Fig. 3.12 is driven by a current of unity into bus 1, the driving point impedance Z_{11} is the sum of Z_1 and Z_2.

$$Z_{11} = 0.00137615 + 0.0077062 = 0.00908257$$

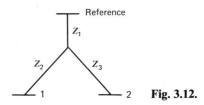

Fig. 3.12.

The voltage of bus 2 is equal to the voltage drop on the impedance $Z_1 = 0.00137615$. This is the transfer impedance Z_{12}.

The driving point impedance of bus 2 is then obtained by driving unit current into bus 2.

$$Z_{22} = Z_1 + Z_3 = 0.00137615 + 0.01155964 = 0.01293579$$

These values agree with the matrix obtained by the Kron reduction. Thus the reduction can be viewed as a network reduction or an algebraic manipulation.

Addition of the Fourth Line Continue the building algorithm by selecting the next line 0–3 for processing. This line is identified to be a type 1 line. The matrix is augmented by a row and column of zeroes and the diagonal of the new axis is the impedance of the line, 0.005. The new bus is added to the bus list.

$$
\text{matrix} \quad
\begin{array}{c}
1 \\
2 \\
3
\end{array}
\begin{bmatrix}
0.00908257 & 0.00137615 & 0 \\
0.00137615 & 0.01293579 & 0 \\
0 & 0 & 0.005
\end{bmatrix}
\quad \text{bus list } 1,2,3
$$

with columns 1 2 3.

Addition of the Fifth Line The next line in the data list (2–3) is a loop closing line, since both buses are in the system bus list. The loop axis is the difference of the columns corresponding to the buses 2 and 3. The diagonal element is obtained from equation 3.12.

$$Z_{\text{loop-loop}} = Z_{22} + Z_{33} - 2Z_{23} + Z_{\text{line } 2\text{-}3}$$

$$= 0.01293578 + 0.005 - (2)(0) + 0.122$$

$$= 0.13993578$$

The augmented matrix is

$$
\begin{array}{c}
1 \\
2 \\
3 \\
\text{loop}
\end{array}
\begin{bmatrix}
0.00908257 & 0.00137615 & 0 & 0.00137615 \\
0.00137615 & 0.01293579 & 0 & 0.01293579 \\
0 & 0 & 0.005 & -0.005 \\
0.00137615 & 0.01293579 & -0.005 & 0.13993579
\end{bmatrix}
$$

with columns 1 2 3 loop.

Reduction of the matrix by application of equation 3.23 gives:

$$
\text{matrix} \quad
\begin{array}{c}
1 \\
2 \\
3
\end{array}
\begin{bmatrix}
0.00906904 & 0.00124893 & 0.00004917 \\
0.00124893 & 0.01173999 & 0.00046220 \\
0.00004917 & 0.00046220 & 0.00482135
\end{bmatrix}
$$

with columns 1 2 3.

bus list $1,2,3$

The system diagram is shown below.

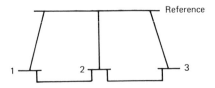

Addition of the Sixth Line The line 2–4 is identified to be a type 2 line, a line from the existing bus 2 to a new bus 4. A new axis is added to the matrix. The diagonal element is determined by use of equation 3.9.

$$Z_{44} = Z_{22} + Z_{\text{line 2-4}} = 0.011739999 + 0.084 = 0.095739999$$

The off diagonal elements are obtained from equation 3.8 in which $q = 4$ and $p = 2$. The row corresponding to bus 4 is identical to the row of bus 2.

	1	2	3	4
1	0.00906904	0.00124893	0.00004917	0.00124893
2	0.00124893	0.01173999	0.00046220	0.01173999
3	0.00004917	0.00046220	0.00482135	0.00046220
4	0.00124893	0.01173999	0.00046220	0.09573999

matrix

bus list 1, 2, 3, 4

The system diagram is shown below.

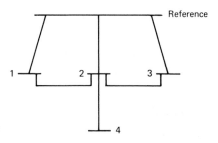

Addition of the Seventh Line Addition of the line 3-5 gives:

$$Z_{55} = Z_{33} + Z_{\text{line 3-5}} = 0.00482135 + 0.037 = 0.04182135$$

Column 5 is a duplicate of column 3.

matrix

	1	2	3	4	5
1	0.00906904	0.00124893	0.00004917	0.00124893	0.00004917
2	0.00124893	0.01173999	0.00046220	0.01173999	0.00046220
3	0.00004917	0.00046220	0.00482135	0.00046220	0.00482135
4	0.00124893	0.01173999	0.00046220	0.09573999	0.00046220
5	0.00004917	0.00046220	0.00482135	0.00046220	0.04182135

bus list 1, 2, 3, 4, 5

Addition of the Eighth Line Addition to the system of line 1-6 which is a line from an existing bus 1 to a new bus 6 is carried out as indicated for line 3-5 in step 7 (type 2 line).

Addition of the Ninth Line The line 6-7 is also a type 2 line in which the line is from existing bus 6 to a new bus 7. The process is illustrated in step 7. After addition of these two lines the matrix is

	1	2	3	4	5	6	7
1	0.00906904	0.00124893	0.00004917	0.00124893	0.00004917	0.00906904	0.00906904
2	0.00124893	0.01173999	0.00046220	0.01173999	0.00046220	0.00124893	0.00124893
3	0.00004917	0.00046220	0.00482135	0.00046220	0.00482135	0.00004917	0.00004917
4	0.00124893	0.01173999	0.00046220	0.09573999	0.00046220	0.00124893	0.00124893
5	0.00004917	0.00046220	0.00482135	0.00046220	0.04182135	0.00004917	0.00004917
6	0.00906904	0.00124893	0.00004917	0.00124893	0.00004917	0.13506904	0.13506904
7	0.00906904	0.00124893	0.00004917	0.00124893	0.00004917	0.13506904	0.30306904

Addition of the Tenth Line The line 4-7 is a loop closing line, since both buses are already in the system. The loop row and column are obtained by taking the difference of rows 4 and 7 and the columns 4 and 7. The diagonal element of this new axis is obtained from equation 3.12.

$$Z_{\text{loop-loop}} = Z_{44} + Z_{77} - 2Z_{47} + Z_{\text{line 4-7}}$$

$$= 0.09573999 + 0.30306904 - 2(0.00124893) + 0.084$$

The loop axis is eliminated as illustrated in steps 3 and 5.

Addition of the Eleventh Line The line 5-8 is determined to be a type 2 line and its addition has been illustrated in step 6.

Addition of the Last line Line 7-8 is loop closing line. The method has been illustrated in step 3. It is readily seen that regardless of the complexity of the network it can be assembled by this simple means of adding one line at a time. The completed matrix of the sample system is given for reference purposes.

	1	2	3	4
1	0.00889104	0.00132842	0.00011167	0.00204268
2	0.00132842	0.01134623	0.00055371	0.00833387
3	0.00011167	0.00055371	0.00475959	0.00120070
4	0.00204268	0.00833387	0.00120070	0.06620613
5	0.00056903	0.00137805	0.00425613	0.00792254
6	0.00626215	0.00303974	0.00085568	0.01834370
7	0.00275695	0.00532152	0.00184769	0.04007839
8	0.00102639	0.00220239	0.00375267	0.01464439

5	6	7	8
0.00056903	0.00626215	0.00275695	0.00102639
0.00137805	0.00303974	0.00532152	0.00220239
0.00425613	0.00085568	0.00184769	0.00375267
0.00792254	0.01834370	0.04007839	0.01464439
0.03662437	0.00652532	0.01446703	0.03199261
0.00652532	0.08999879	0.03364765	0.01219496
0.01446703	0.03364765	0.07483526	0.02708638
0.03199261	0.01219496	0.02708638	0.06023255

FAULT ANALYSIS OF A SYSTEM

A complete analysis of the system is possible once the Z-matrix of the system is completed. Using values that are in the matrix of the sample network for illustrative purposes, consider node 3 to be in fault condition. The matrix element $Z_{33} = 0.00475959$ means that, if a voltage of that value is applied between bus 3 and the reference, a total current of 1.0 will flow through the network. The full voltage of the generator will cause a current that can be determined by considering

$$\frac{I'}{I} = \frac{E'}{E} \qquad (3.24)$$

in which $I = 1.0$ when $E = 0.00475959$. It is desired to know I' when $E' = 1.0$.

$$I' = \frac{1.0}{0.00475959} = 210.10 \text{ pu}$$

This is the result that is obtained by consideration of the matrix 3.2 and the equation 3.3.

The total fault value can be obtained either in amperes, pu, or MVA by dividing the base amperes, unity, or pu MVA base of the line data by the corresponding diagonal element of the matrix (see equation 3.3).

The contribution to the fault by a line is computed using equation 3.5. The contribution from bus 2 to the fault is

$$I_{23} = \frac{Z_{33} - Z_{32}}{Z_{\text{line 2-3}}} \frac{1.0}{Z_{33}} = (0.00475959 - 0.00055371)/(0.122)(0.00475959)$$

$$= 7.2 \text{ pu}$$

Note: The base here was 1.0 and the flow is pu. Base amperes or MVA could also have been used.

VOLTAGES ON BUSES DURING FAULT CONDITION

The voltage on bus 2, when unit current is injected into bus 3, is $Z_{23} = 0.00055371$. The voltage on bus 3 at this time is 0.00475959. The difference in voltage is $Z_{33} - Z_{32} = 0.00475959 - 0.00055371 = 0.00420588$. This difference occurs if the current is 1.0 but the fault current was determined to be $1/Z_{33}$. The difference in voltage between bus 2 and 3 is therefore

$$\frac{Z_{33} - Z_{32}}{Z_{33}} = \frac{0.00420588}{0.00475959} = 0.8835 \text{ pu}$$

but the voltage of bus 3 under fault condition is zero. Therefore, the voltage of bus $2 = 0.8835$.

OPENING A LINE DURING A STUDY

A line of a system may be opened or removed by adding a line in parallel with the existing line. The impedance of the new line to be added is the negative of the original line. The loop closing equations 3.12 and 3.13, and the elimination of the loop by a Kron reduction 3.15 are used.

In the course of a complete fault study it is often desirable to open each line connected to a faulted bus, one at a time, and to obtain the new total fault and the contribution of the remaining lines. It is undesirable to modify the matrix of the total system because a great deal of computation would be done unnecessarily on elements that are not required in the analysis. Furthermore, it is undesirable (because of rounding) to remove a line, add it back, remove another, add it back, and so on. Errors would accumulate in the Z-matrix elements because of the repetative modification of the matrix.

The best method is to extract a small matrix, from the total matrix, that includes the driving point and transfer impedances of the bus to be faulted and its immediate neighbors.

For example, if bus 3 is to be faulted, extract the small matrix

	3	2	5
3	0.00475959	0.00055371	0.00425613
2	0.00055371	0.01134623	0.00137805
5	0.00425613	0.00137805	0.03662437

To open the line 3-5, add a line (loop closing) whose impedance is the negative of the original line impedance -0.037. The diagonal element of the loop axis is

$$Z_{\text{loop-loop}} = Z_{33} + Z_{55} - 2Z_{35} + Z_{\text{line 3-5}}$$

$$= 0.00475959 + 0.03662437 - 2(0.00425613) - 0.037 = -0.00412830$$

The loop column is obtained by subtracting column 5 from column 3.

$$
\begin{array}{c}
 \\
3 \\
2 \\
5 \\
\text{loop}
\end{array}
\begin{array}{cccc}
3 & 2 & 5 & \text{loop} \\
\left[\begin{array}{cccc}
0.00475959 & 0.00055371 & 0.00425613 & 0.00050346 \\
0.00055371 & 0.01134623 & 0.00137805 & -0.00082434 \\
0.00425613 & 0.00137805 & 0.03662437 & -0.03236824 \\
0.00050346 & -0.00082434 & -0.03236824 & -0.00412830
\end{array}\right]
\end{array}
$$

In the interest of efficiency the Kron reduction is used to modify only row 3, since a fault on 3 can be completely analyzed with only these values.

The modified row vector that reflects the opening of the line 3-5 is

$$
\begin{array}{cccc}
 & 3 & 2 & 5 \\
3 & [0.00482099 & 0.00045317 & 0.00030870]
\end{array}
\tag{3.25}
$$

The new total fault is

$$
\frac{\text{base}}{Z_{33}} = \frac{1}{0.00482099} = 207.43 \text{ pu}
$$

The contribution from bus 2 is found to be 7.4 pu. The flow from bus 5 to 3 over the line $X = 0.037$ is the same magnitude as the flow over the line $X = -0.037$, but of opposite sign. The net contribution from 5 to 3 is therefore zero.

Having the row vector, which gives the values for the fault on bus 3 with the line 3-5 open, affords an excellent opportunity to determine the magnitude of the fault current that would flow if the fault was removed from bus 3 and placed on the line side of the open breaker at bus 5 on the 3-5 line (see Fig. 3.13).

The row vector gives the driving point and transfer impedances of bus 3 with the line open. The driving point impedance of bus 9 is obtained from equation 3.9. Since the line 3-9 is a radial from bus 3

$$
Z_{qq} = Z_{33} + Z_{\text{line 3-5}} = 0.00482099 + 0.037 = 0.04182099
$$

The total fault at point $9 = \text{base}/0.04182099 = -23.91$ pu.

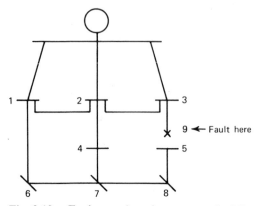

Fig. 3.13.   Fault transferred to open end of line.

BUS TIE BREAKERS

Certain power systems have bus tie breakers that may be closed or opened for operating reasons. For example, a generating station may have two generators connected to a bus system as shown in Fig. 3.14.

If either generator is out of service for overhaul, or for some operating reason, it may be necessary to close the bus tie breaker. The matrix can be modified to represent the condition of the closed tie breaker by adding a line of zero impedance between bus P and Q. This is a perfectly satisfactory method of modeling the system for the closed bus tie breaker, but it would be impossible to open the closed breaker later in the study. This becomes evident by considering the addition of a line of -0.0 impedance in an attempt to open the line of impedance of 0.0 corresponding to the closed breaker. When the breaker is closed, the two columns (P and Q)

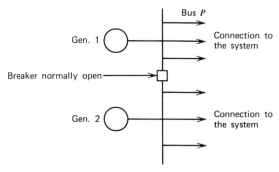

Fig. 3.14.

become identical in value, since there is zero impedance connected between the buses. The diagonal element for the addition of the -0.0 impedance element, in parallel with the 0.0 impedance element, in an attempt to open the breaker becomes

$$Z_{\text{loop-loop}} = Z_{pp} + Z_{qq} - 2Z_{pq} + (-0.0)$$

but

$$Z_{pp} = Z_{qq} = Z_{pq}$$

when the buses are tied and $Z_{\text{loop-loop}} = 0$.

Substitution of this value in equation 3.23 is not permitted. Therefore, the bus tie breaker can not be opened if this representation is used.

BUS TIE BREAKER THAT CAN BE OPENED

To tie two buses with a bus tie impedance of zero that can be opened one introduces a fictitious bus between the buses to be tied and adds a line from P to T with an impedance of Z_1 and a line from T to Q with an impedance of $-Z_1$ (see Fig. 3.15).

The impedance from bus P to bus Q is zero but now the breaker may be opened by adding a line from bus T to bus Q with impedance $+Z_1$. This removes one side of the bus tie breaker.

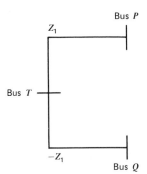

Fig. 3.15. Simulation of a closed bus tie breaker.

LIMITATION OF THE Z-MATRIX METHOD

Since the Z-matrix is full, that is, since there is a transfer impedance for every bus with respect to every other bus, the memory requirement for storage of the Z-matrix is very great. Because the X to R ratio of high voltage systems is high (of order 10 for example) the Z-matrix can be computed using X only without seriously affecting the accuracy of a study. Furthermore, the matrix is symmetric and can therefore be stored in upper triangular form. A system of 100 buses will require a matrix of 5050 elements $N(N+1)/2$. A system of 200 buses has 20100 elements. As late as 1958 fault studies of major power systems were made on analog models with fewer than 100 buses. The introduction of the digital computer programs increased very rapidly the number of buses used in the representation. It became necessary to devise methods to increase the program size. This aspect of the problem is discussed in Chapter 4.

References

1. **J. B. Ward and H. W. Hale**, Digital solution of power flow problems, *Trans. AIEE*, Vol. 75, Part III, (1956), pp. 398–404.
2. **L. W. Coombe and D. G. Lewis**, Digital calculations of short circuit currents in large complex-impedance networks, *Trans. AIEE*, Vol. 75, Part. III, (1956), pp. 1394–1397.
3. **H. E. Brown, C. E. Person, L. K. Kirchmayer, and G. W. Stagg**, Digital calculation of three-phase short circuits by matrix method, *Trans. AIEE*, Vol. 79, Part III, (1960), pp. 1277–1281.
4. **G. Kron**, *Tensor Analysis of Networks*, Wiley, 1939.
5. **A. H. El-Abiad**, Digital calculation of line-to-ground short circuits by matrix methods, *Trans. AIEE*, Vol. 79, Part III, (1960), p. 323.
6. **R. A. Frazer, W. J. Duncan, and A. R. Collar**, *Elementary Matrices*, Cambridge Press, 1955.

4

Extension of the Z-Matrix Short-Circuit Program to Include Very Large Systems

Originally the proper sequencing of line data for processing by the Z-matrix building algorithm could be done manually because of the limited amount of data required to describe a network of 100 buses and 300 lines. This was the maximum size of system that could be studied in 1958 using this method. This limitation was imposed by the size of the matrix that could be stored in the 8K word memory of the computers available at that time. The success of the Z-matrix short-circuit method [1] and the availability of computers with 32K memories extended the program capability to systems of 220 nodes in 1959. For systems of this size, the proper sequencing of the line data for processing by the building algorithm was no longer suited to manual methods.

A sorting routine was developed which reordered a randomly ordered data list such that the data could be processed sequentially by the matrix building algorithm. The list was not reordered optimally, but as each line was selected from the reordered line list, it ensured that it would be possible to connect it to the network that had been assembled. A technique was later developed by Baumann [2] that optimally ordered the line data to minimize the time required for the building algorithm. Subsequently an axis discarding technique was developed that permitted very large systems to be studied. This discarding technique [3] eliminated the need for the optimal ordering method.

Short-circuit studies of 220-bus systems created the demand for an equivalent program to carry out studies in even greater detail. An equivalent program was written that would reduce a 200-bus 300-line portion of a system to an equivalent mesh network between a few retained buses. Thus a system of 400 buses could be split in any convenient manner into two

49

nearly equal parts. One portion could be reduced to an equivalent, which would then be connected to the unreduced part. The short-circuit analysis of the unreduced portion is completed and the equivalent properly accounts for the part that had been reduced.

The equivalent program encouraged the representation of systems in even greater detail by multiple use of the technique in the reduction of portions of the system to mesh equivalents. To study a system of 1000 buses would require reduction of the system to several mesh equivalents. The preparation of the data for a study of the entire system would necessitate manually combining the various permutations of equivalent networks with the detailed representation of single areas. This was inefficient and the excessive card handling resulted in many data errors. An automatic method of studying large systems (in excess of 2500 buses) by an axis discarding technique was developed [3] that enabled large systems to be studied without resorting to the use of equivalents.

The automatic line sequencing for the matrix building algorithm, reduction of a portion of a system to an equivalent, and the axis discarding technique are discussed in this chapter.

DATA PREPARATION

Numbers are assigned arbitrarily to the nodes of the system with the exception of the reference node which is assigned the number zero. (In short-circuit work it is convenient to make the common bus behind the generator reactances the reference bus). Here node or bus is used to describe any junction point in the transmission system of two or more elements of the network. Node is an inclusive term, since it includes not only buses as used in power system work, but any junction point where two or more elements of the network are joined together. For instance a tap point on a transmission line or a center point of a transformer equivalent is a node and must be given a unique number. The elements of the network (transmission line, transformer winding, generator internal reactance, etc.) are described by the node numbers at the ends of the element, and the impedances of the element in the positive and zero sequence networks [4]. The impedance in both networks is required when single phase short circuits are considered in Chapter 5.

The positive and zero sequence networks are made to correspond element by element. A one-to-one correspondence is achieved by adding an element of infinite impedance where an element is missing in one of the networks. This technique is illustrated by the network representation of a generator connected to the system by a $\Delta - Y$ transformer in which the neutral of the Y is grounded (see Fig. 4.1).

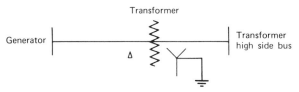

Fig. 4.1

The generator is a source in the positive sequence network but not in the zero sequence. The transformer ground is a source in the zero sequence but not in the positive sequence network. The generator, ground, and high side bus are represented by a 4-bus equivalent (see Fig. 4.2). A fictitious node is created in the interior of the transformer and half of the impedance of the transformer is placed on both sides of this node.

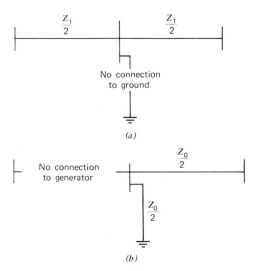

Fig. 4.2. (*a*) Positive sequence network. (*b*) Zero sequence network.

The two diagrams are made to correspond by addition of elements with infinite impedances (see Fig. 4.3). The data describing this portion of the network is given in Table 4.1.

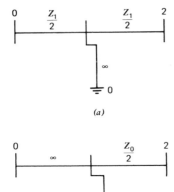

(a)

(b)

Fig. 4.3. (a) Positive sequence network.
(b) Zero sequence network.

Table 4.1

Line		Impedances	
		Positive	
Node	Node	Sequence	Zero Sequence
0	1	$Z_1/2$	∞
0	1	∞	$Z_0/2$
1	2	$Z_1/2$	$Z_0/2$

DETAILS OF THE LINE REORDERING TECHNIQUE

The sorting routine reorders a randomly ordered set of line data into a feasible ordering of the lines, which will permit the matrix formation algorithm to process the line data sequentially. The reordered line list is not optimally ordered but is a feasible ordering that is dependent on the original random order of the line data. The details of the sorting operation follow:

1. The line data are examined for a line connected to the reference. The

common bus behind the generator internal reactances is the reference bus in short-circuit study work. The first line that is found connected to the reference is transferred to the reordered line list. The other node at the end of this line determines the first node in the system that is assembled from the line data.

2. The line data are examined and the nodes connected directly to the first node are added to the node list of the system that is assembled.

3. Advance to the next node of the list being assembled. Examine all lines in the unordered list for lines connected to this node.

a. Lines connected between this node and the reference node are added to the reordered line list.

b. Lines between it and nodes that preceed it in the list being assembled are added to the reordered line list.

c. Lines between it and nodes that succeed it in the list are passed over for later processing.

d. Nodes are added to the list for those lines between the node being processed and nodes that do not already occur in the list. The line is not processed at this time.

4. After all lines have been examined return to step 3 and repeat.

In case the system being studied is made up of several disconnected areas (no node in common except the reference node), lines remain in the unordered line list after processing the last node in the list of nodes of the system that is being assembled. The lines remaining in the randomly ordered data list are examined for another line connected to the reference that could be used for restarting the process. If one is found, the other node that describes the line is added to the node list; the line is added to the reordered line list. Return to step 3 and continue. After all possible lines have been reordered, lines may remain in the unordered line list. It must then be concluded that there is a data error, since the remaining lines are disconnected from the system and from the reference. The remaining line data should be printed for analysis so that the data error might be more readily corrected. The arrangement of data and a flow chart of this routine are given at the end of the chapter.

EXAMPLE OF THE LINE REORDERING ROUTINE

Consider the line data (slightly rearranged) that is used in Chapter 3 to illustrate the building algorithm.

Line	X(pu)
1-2	0.084
0-1	0.010
0-2	0.015
0-3	0.005
2-3	0.122
4-2	0.084
3-5	0.037
1-6	0.126
7-6	0.168
4-7	0.084
8-5	0.037
7-8	0.140

Application of step 1 to the line data results in the following:

Reordered Line List	System Bus List
0-1	1

Application of step 2 expands the bus list to the following.

Reordered Line List	System Bus List
0-1	1 2 6

↑
Bus being processed

Application of step 3 expands the line list by step a and b. The bus list is expanded by step d.

Line List	Bus List
0-1	1 2 6 3 4
1-2	↑
0-2	Bus being processed

Continuation of the process results in the following reordered data and system bus list.

Line Data	Bus List
0-1	1 2 6 3 4 7 5 8
1-2	
0-2	
1-6	
0-3	
2-3	
4-2	
7-6	
4-7	
3-5	
8-5	
7-8	

PROCESSING LINES OF INFINITE IMPEDANCE

Loop Closing Lines   The addition of a loop closing line that has infinite impedance results in a degenerate matrix modification as can be seen by consideration of equations 3.12 and 3.23 of Chapter 3. Application of equation 3.12 when $Z_{\text{line } p-q} = \infty$ yields:

$$Z_{\text{loop-loop}} = Z_{pp} + Z_{qq} - 2Z_{pq} + Z_{\text{line } p-q} = \infty$$

Use of this value in equation 3.23 for modification of the elements of the matrix when the loop axis is eliminated by the Kron reduction leaves all elements of the matrix unchanged.

$$Z'_{ij} = Z_{ij} - Z_{i\text{-loop}}\left(\frac{1}{\infty}\right)Z_{\text{loop-}j} = Z_{ij}$$

Therefore in the matrix building algorithm, if a loop closing line is encountered that has infinite impedance, the matrix remains unchanged, and all calculations indicated by equations 3.12, 3.13, and 3.23 are omitted.

A Line from the Reference to a New Bus   The addition of a line of infinite impedance from the reference bus to a new bus can lead to complications as can be seen by processing the line data for the zero sequence network from Table 4.1. Assume that the Z_0 matrix already has several axis at the time the three lines of Table 4.1 are encountered (see Fig. 4.4).

Fig. 4.4.   Matrix before encountering a radial line of infinite impedance.

Addition of the first line 0-1 to the system by application of equations 3.6 and 3.7 results in a matrix as shown in Fig. 4.5.

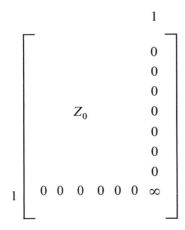

Fig. 4.5.   Matrix after processing an infinite impedance line to the reference.

Addition of the next line 0-1 whose impedance is $Z_0/2$ by applying equations 3.12 and 3.23 gives

$$Z_{\text{loop-loop}} = Z_{1\text{-}1} + Z_{0\text{-}0} - 2Z_{0\text{-}1} + Z_{\text{line}\,0\text{-}1} \tag{4.1}$$

It must be remembered that the driving point impedance of the reference bus Z_{0-0} is zero, and also that the transfer impedance of all buses with respect to the reference bus are also zero (see Fig. 4.6). The voltage of the reference bus with respect to itself, when current is injected into any bus of the system, is zero.

Equation 4.1 gives

$$Z_{\text{loop-loop}} = \infty + 0 - (2.0)(0) + \frac{Z_0}{2} = \infty$$

Equation 3.13 degenerates into the following:

$$Z_{\text{loop-}i} = Z_{1-i} - Z_{0-i} = 0\text{-}0 = 0$$

and

$$Z_{i\text{-loop}} = Z_{i-1} - Z_{i-0} = 0\text{-}0 = 0$$

The matrix before reduction is

$$
\begin{array}{cc}
 & \begin{array}{cc} 1 & \text{loop} \end{array} \\
\left[
\begin{array}{ccccccc|cc}
 & & & & & & & 0 & 0 \\
 & & & & & & & 0 & 0 \\
 & & & & & & & 0 & 0 \\
 & & & Z_0 & & & & 0 & 0 \\
 & & & & & & & 0 & 0 \\
 & & & & & & & 0 & 0 \\
 & & & & & & & 0 & 0 \\
0 & 0 & 0 & 0 & 0 & 0 & 0 & \infty & \infty \\
0 & 0 & 0 & 0 & 0 & 0 & 0 & \infty & \infty \\
\end{array}
\right]
\end{array}
\begin{array}{c} \\ \\ \\ \\ \\ \\ \\ 1 \\ \text{loop} \end{array}
$$

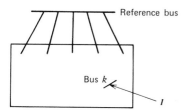

Fig. 4.6. The reference bus remains at zero potential for all current injections.

Reduction of the matrix by use of equation 3.23 eliminates the loop axis. All of the matrix of Fig. 4.5 remains unchanged with the exception of the element Z_{11} which is indeterminant:

$$Z_{11} = \infty - \frac{(\infty)(\infty)}{\infty}$$

The correct value of this diagonal element is $Z_0/2$ as can be seen from Fig. 4.7 in which the two impedances ∞ and $Z_0/2$ are connected between the reference bus and bus 1. This correct value would be obtained if the two lines had been processed in reverse order. Verification of this is left as an exercise for the student. (Rember to apply the technique of loop closing of lines with infinite impedance.)

The identical difficulty is encountered when adding a line from a bus of the system to a new bus. This type of line will not be discussed, since a single remedy will eliminate the difficulty. To ensure that the matrix contains the proper values of driving point and transfer impedance when radial lines with infinite impedance are encountered the building algorithm is modified.

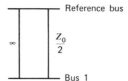

Fig. 4.7.   Parallel equivalent.

MODIFICATION OF THE BUILDING ALGORITHM

When it has been determined that a new axis is required in the matrix because the line to be added is a line from the reference to a new bus or a line from an existing bus to a new bus, a fictitious line is processed first. The fictitious line is a line from the reference to the new bus with any finite value within the range of the system line data. This guarantees that the new diagonal is well defined. All system lines connected to the new bus are then loop closing lines and are processed by equations 3.12, 3.13, and 3.23 for lines of finite impedance or by the method discussed for lines of infinite impedance. When all lines connected to the bus have been processed, the diagonal element of its axis is examined. If the diagonal element contains the value of the impedance of the fictitious line, it indicates that all lines connected to the bus were infinite and no processing has been done. The driving point impedance of this bus in this sequence network is therefore truly infinite and the diagonal element is replaced by infinity. If the value of the diagonal element is not equal to the impedance of the fictitious line, it indicates that the bus does have finite connections other than the fictitious line. An additional fictitious line is added to the system between the reference and the node. The impedance of this second fictitious line is equal to the negative of the first fictitious line. This final line addition removes the original fictitious line and restores the matrix elements to their proper values.

REDUCTION OF A SYSTEM TO AN EQUIVALENT

The purpose of a mesh equivalent is to eliminate unnecessary nodes from a network to permit a given computer to be used to study larger systems than would otherwise be possible. A mesh equivalent must have the same apparent impedance between the nodes that are not eliminated as does the original network. To accomplish what is intended, simplification of the study, the number of the retained nodes should be limited to less than 30, otherwise the number of lines in the equivalent mesh is excessive. By eliminating equivalent lines of very high impedance, an approximate equivalent can be obtained with a much greater number of retained buses. A mesh equivalent is interconnected with an unreduced portion of the system being studied and the study greatly simplified by the reduction of the number of buses represented.

A power system has only a few lines (perhaps three is an average) connected to each bus. A mesh on the other hand has a line from every node to every other node. An N node equivalent will therefore have $(N^2 - N)/2$ total lines. Engineering judgment must be exercised to keep the number of nodes retained in the equivalent as small as possible so that the lines in the equivalent will not become a burden in subsequent calculations.

COMPUTATION OF AN EQUIVALENT

The Z-matrix of a portion of the system that is to be reduced is formed by the building algorithm (see Chapter 3). A small matrix Z_s with axes corresponding to only those buses to be retained, is extracted (no mathematical procedure required) from the complete matrix. This small matrix gives the driving point and transfer impedance of the original network for the retained nodes with respect to the reference node. The small matrix is inverted to obtain the nodal admittance matrix of the equivalent mesh network between the retained nodes. It should be remembered that a diagonal element y_{ii} of the nodal admittance matrix is the sum of the admittances connected to the particular node i. The off-diagonal element y_{ij} is the negative of the admittance of the connection between the nodes i and j. Furthermore, the reference node does not have a corresponding axis in the matrix but the admittances of lines connected to the reference node are included in the diagonal element summations.

$$Z_s^{-1} = Y_s = \begin{bmatrix} \Sigma y_1 & -y_{12} & -y_{13} & \cdots \\ -y_{21} & \Sigma y_2 & -y_{23} & \cdots \\ \cdot & \cdot & \cdot & \cdots \\ \cdot & \cdot & \cdot & \cdots \end{bmatrix}$$

Define a new matrix Y' in which the diagonal elements of the matrix Y_s are replaced by the negative of the sum of the elements of the particular row. Since

$$y_{ii} = y_{o-i} + \sum_{\substack{j=1 \\ j \neq i}}^{n} y_{ij}$$

and for each element in the summation on the right there is a corresponding negative off-diagonal term, the sum of the elements of row i is equal to y_{o-i} the admittance of the tie to the reference. Each diagonal element of the matrix Y' is the negative of the sum of the elements of its row and is equal to $-y_{o-i}$. All elements in Y' are the negative admittances of the connections between the various nodes, including the reference node. The admittances of ties to the reference are stored as the diagonal elements.

Define a new matrix Z' in which each element is the negative reciprocal of each element of Y', element by element. The elements of Z' are the primitive impedances of lines in the mesh equivalent.

$$Z'_{ij} = Z_{\text{line } ij} \qquad Z'_{ii} = Z_{\text{line } 0-i}$$

Example   Obtain the equivalent of the sample network of Fig. 3.8 as it would appear if viewed looking into buses 3 and 8.

From the complete matrix of the 8-bus system of Ref. 1 extract the corresponding small Z_s matrix (see p. 42 Chapter 3).

$$Z_s = \begin{matrix} \\ 3 \\ 8 \end{matrix} \begin{matrix} \quad 3 \qquad\qquad 8 \\ \begin{bmatrix} 0.00475959 & 0.00375267 \\ 0.00375267 & 0.06023255 \end{bmatrix} \end{matrix}$$

Inversion of this small matrix gives the Y-matrix.

$$
Z_s^{-1} = Y_s = \begin{array}{c} \\ 3 \\ 8 \end{array} \begin{array}{cc} 3 & 8 \\ \left[\begin{array}{cc} 220.95603 & -13.76623 \\ \\ -13.76623 & 17.46000 \end{array}\right] \end{array}
$$

Summing each row and replacing the diagonal by the negative of the sum gives the Y'-matrix which was just defined.

$$
Y' = \begin{array}{c} \\ 3 \\ 8 \end{array} \begin{array}{cc} 3 & 8 \\ \left[\begin{array}{cc} -207.18980 & -13.76623 \\ \\ -13.76623 & -3.69377 \end{array}\right] \end{array}
$$

Obtaining the negative reciprocals term by term:

$$
Z' = \begin{array}{c} \\ 3 \\ 8 \end{array} \begin{array}{cc} 3 & 8 \\ \left[\begin{array}{cc} 0.0048265 & 0.07264153 \\ \\ 0.07264153 & 0.270726 \end{array}\right] \end{array}
$$

The equivalent network is given in Fig. 4.8.

The line data corresponding to the reduced network is as follows:

Line	Reactance
0-3	0.004827
0-8	0.270726
3-8	0.072642

To verify that these lines produce the same driving point and transfer impedance matrix as the original line data, the Z-matrix is formed using these line data.

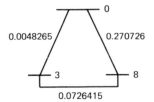

Fig. 4.8

After adding the first two lines the matrix is

$$
\begin{array}{c}
 & \begin{array}{cc} 3 & \quad\quad 8 \end{array} \\
\begin{array}{c} 3 \\ 8 \end{array} &
\left[\begin{array}{cc}
0.004827 & 0 \\
0 & 0.270726
\end{array} \right]
\end{array}
$$

Addition of the loop closing line gives

$$
\begin{array}{c}
 & \begin{array}{ccc} 3 & \quad\quad 8 & \quad\quad \text{loop} \end{array} \\
\begin{array}{c} 3 \\ 8 \\ \text{loop} \end{array} &
\left[\begin{array}{ccc}
0.004827 & 0 & 0.004827 \\
0 & 0.270726 & -0.270726 \\
0.004827 & -0.270726 & 0.348195
\end{array} \right]
\end{array}
$$

$$X_{11} = 0.004827 - (0.004827)(0.004827)/0.348195$$

$$= 0.004827 - 0.000067 = 0.004760$$

$$X_{12} = 0.003753$$

$$X_{22} = 0.060233$$

$$
\left[\begin{array}{cc}
0.004760 & 0.003753 \\
0.003753 & 0.060233
\end{array} \right]
$$

This verifies that the equivalent reduced network gives the same Z-matrix as the original network.

EXTENSION OF THE METHOD TO LARGE SYSTEMS

Assume that it is required to make a short-circuit study of a system whose Z-matrix can not be contained in the core memory of the computer that is available. Let it further be assumed that the matrix of half of the system can be stored in the computer memory. This second restriction will be removed later.

The system can be studied by dividing it in half in some convenient manner (see Fig. 4.9). Form the matrix of part A of the system, including the lines $(A_1 - B_1)$, $(A_2 - B_2)$, $(A_3 - B_3)$ by the matrix building algorithm [1]. The mesh equivalent of part A as it appears when viewed from nodes B_1, B_2, and B_3 is computed by the method "mesh equivalents" discussed earlier in this chapter.

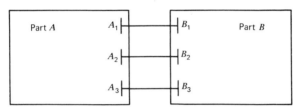

Fig. 4.9.   System divided for application of network reduction.

The lines of the mesh equivalent that are obtained from the Z'-matrix (in the case of a 4-bus equivalent there are six lines) are included with the data of part B (see Fig. 4.10).

The matrix of part B (including the equivalent lines) is formed by the building algorithm. The study of area B can now be completed with area A properly taken into account.

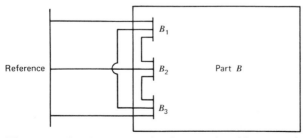

Fig. 4.10.   The system in which Part A has been replaced by an equivalent.

In the formation of the matrix of area B, if the mesh equivalent lines of area A are processed first, the matrix that will result after processing only the equivalent lines will be exactly the matrix Z_s. All of the steps after extraction of the small matrix Z_s (inversion, replacing the diagonals by the negative of the sum of the rows, reciprocation of the elements of the matrix Y', and reforming the matrix corresponding to the lines of the equivalent) are unnecessary. All that was accomplished by these operations was to collect the elements of the small matrix and move them to the upper left-hand corner of the matrix. The remainder of the matrix was erased (see Fig. 4.11).

A further simplification is made in the method which permits part A, the part to be reduced, to no longer be limited in size to a system whose matrix

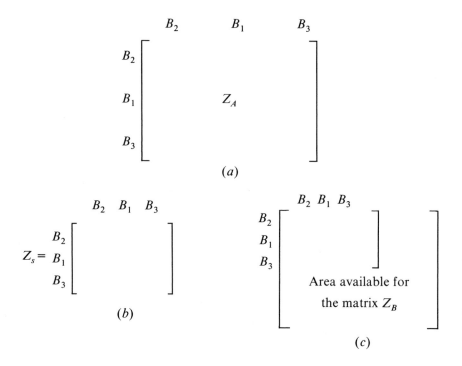

Fig. 4.11.   The steps involved in modifying the matrix for the elimination of area A.

(a) Z-matrix of area A with axes of boundary buses indicated.
(b) The small extracted matrix.
(c) Starting point for forming matrix Z_B.

will fit into the computer high-speed memory.

In the addition of a radial line from an existing bus to a new bus by the building algorithm the only elements that take an active part in the matrix modification are the elements of the row and column of the existing bus (see equations 3.8 and 3.9). In the addition of a loop closing line the only columns that take an active part in the matrix modifications are the columns being connected by the new line (see equations 3.12 and 3.13). Therefore as soon as all lines have been connected to a bus, the elements of the matrix in the row and column corresponding to the bus will never be required as active participants in the matrix modification. If the bus that has no more lines connected to it is not a bus in the area to be studied, the matrix elements will not be used in the fault analysis. The line data of area A include the tie lines to area B. As soon as all lines are connected to a bus of area A the axis corresponding to the bus may be eliminated. The axis is not required in further matrix modifications and it is not required in the fault analysis. An axis of the Z-matrix is eliminated merely by erasing the row and column and closing up the matrix. The least amount of movement of matrix elements in eliminating a matrix row and column is achieved by replacing the axis no longer required by the last axis of the matrix and erasing the last row and column.

THE LARGE SHORT-CIRCUIT PROGRAM

The data are submitted for a large system. The bus numbers of an area to be studied are supplied by the investigator. The program then operates as follows:

1. The data are sorted. Lines described by buses, both of which are in the area of study, are placed in one file. All other lines are placed in a file of lines in the outside area.

2. The data outside the area of study are sorted by the sorting routine as described earlier.

It may happen that area A is composed of several disconnected areas. In that case when all of the data in one subarea are completed, a new line to the reference is found and the process continued.

It can happen that a subarea of area A remains, which is isolated from the remainder of area A and does not have a reference line; then the process can not be continued. In that case the remaining data of area A are combined with the data of the area B and the line sorting goes on to completion. The data are now ready for processing by the building algorithm.

Example The line data of the 8-bus problem are given. It is desired to stud⸱⸱ only the area defined by buses 6, 7, and 8. The data has been arb⸱ arily rearranged into a random sequence for illustration purposes.

Input Line List

Line	X (pu)
0-3	0.005
0-1	0.01
0-2	0.015
2-4	0.084
3-5	0.037
1-2	0.084
2-3	0.122
1-6	0.126
6-7	0.168
4-7	0.084
5-8	0.037
7-8	0.140

The result of the several line sortings produces a feasible line list that will permit the matrix to be formed. The steps taken in the line sorting can be easily followed and the following sorted line list is obtained.

Line List after Reordering

Line		X
0-3		0.005
3-5		0.037
0-2		0.015
3-2		0.122
5-8	area A	0.037
2-4		0.084
0-1		0.010
2-1		0.084
4-7		0.084
1-6		0.126
8-7	area B	0.140
7-6		0.168

3

Processing the first line of the system, the matrix is 3[0.005] and the retained bus list has bus 3 as its only entry.

After adding the next line of the reordered line list to the system, the matrix is saved in upper triangular form as

$$
\begin{array}{cc}
3 & 5 \\
3 \\
5
\end{array}
\left[
\begin{array}{cc}
0.005 & 0.005 \\
& 0.042
\end{array}
\right]
$$

The bus list contains 3, 5. Addition of the third line to the system produces the matrix

$$
\begin{array}{ccc}
3 & 5 & 2 \\
3 \\
5 \\
2
\end{array}
\left[
\begin{array}{ccc}
0.005 & 0.005 & 0 \\
& 0.042 & 0 \\
& & 0.015
\end{array}
\right]
\quad \text{bus list 3, 5, 2}
$$

Addition of the next line to the system completes the lines to bus 3, which is not in the area of study, and therefore the corresponding matrix elements need not be saved.

After the matrix has been modified to reflect the addition of this line, the first row and column are replaced by the last row and column. Bus 3 is deleted from the bus list and the following matrix results:

$$
\begin{array}{cc}
2 & 5 \\
2 \\
5
\end{array}
\left[
\begin{array}{cc}
0.0134155 & 0.00052817 \\
& 0.04182394
\end{array}
\right]
\quad \text{Bus list 2, 5}
$$

Addition of the line 5-8 completes the lines to bus 5 which is not in the area of study. The matrix elements are modified. The matrix elements for bus 8 replace those of bus 5; bus 5 is deleted from the bus list. The matrix is

$$
\begin{array}{c}
\begin{array}{cc} \quad\quad 2 \quad\quad\quad\quad 8 \end{array} \\
\begin{array}{c} 2 \\ 8 \end{array}
\left[
\begin{array}{cc}
0.0134155 & 0.00052817 \\
 & 0.07882394
\end{array}
\right]
\end{array}
\quad \text{Bus list 2, 8}
$$

Addition of the radial line 2 to 4 produces the matrix

$$
\begin{array}{c}
\begin{array}{ccc} \quad\quad 2 \quad\quad\quad\quad\quad 8 \quad\quad\quad\quad\quad 4 \end{array} \\
\begin{array}{c} 2 \\ 8 \\ 4 \end{array}
\left[
\begin{array}{ccc}
0.0134155 & 0.00052817 & 0.0134155 \\
 & 0.07882394 & 0.00052817 \\
 & & 0.0974155
\end{array}
\right]
\end{array}
\quad \text{Bus list 2, 8, 4}
$$

Addition of the line 0 to 1, a line from the reference to a new bus, gives the matrix

$$
\begin{array}{c}
\begin{array}{cccc} \quad 2 \quad\quad\quad\quad 8 \quad\quad\quad\quad 4 \quad\quad\quad 1 \end{array} \\
\begin{array}{c} 2 \\ 8 \\ 4 \\ 1 \end{array}
\left[
\begin{array}{cccc}
0.0134155 & 0.00052817 & 0.0134155 & 0 \\
 & 0.07882394 & 0.00052817 & 0 \\
 & & 0.0974155 & 0 \\
 & & & 0.010
\end{array}
\right]
\end{array}
\quad \text{Bus list 2, 8, 4, 1}
$$

Addition of the loop closing line 1 to 2 completes the lines to bus 2. After the matrix is modified, the matrix elements of bus 1 replace those of bus 2. Bus 2 is deleted from the bus list. The matrix is

$$
\begin{array}{c}
\begin{array}{ccc} \quad\quad 1 \quad\quad\quad\quad\quad 8 \quad\quad\quad\quad\quad 4 \end{array} \\
\begin{array}{c} 1 \\ 8 \\ 4 \end{array}
\left[
\begin{array}{ccc}
0.00906904 & 0.00004917 & 0.00124893 \\
 & 0.07882134 & 0.00046220 \\
 & & 0.0957400
\end{array}
\right]
\end{array}
\quad \text{Bus list 1, 8, 4}
$$

Addition of the line 4 to 7 completes the lines to bus 4. The matrix is modified and then the matrix elements for bus 7 replace those of bus 4.

Bus 4 is removed from the bus list. The matrix is

$$
\begin{array}{c}
 & 1 & 8 & 7 \\
\begin{array}{c} 1 \\ 8 \\ 7 \end{array}
\left[
\begin{array}{ccc}
0.00906904 & 0.00004917 & 0.00124893 \\
 & 0.07882134 & 0.00046220 \\
 & & 0.17973998
\end{array}
\right]
\end{array}
\qquad \text{bus list } 1, 8, 7
$$

The addition of the radial line 1 to 6 completes the lines of bus 1. The matrix elements of bus 6 replace those of bus 1. Bus 1 is removed from the bus list. The matrix is

$$
\begin{array}{c}
 & 6 & 8 & 7 \\
\begin{array}{c} 6 \\ 8 \\ 7 \end{array}
\left[
\begin{array}{ccc}
0.13506903 & 0.00004917 & 0.00124893 \\
 & 0.07882134 & 0.00046220 \\
 & & 0.17973998
\end{array}
\right]
\end{array}
\qquad \text{bus list } 6, 8, 7
$$

The two remaining lines are in the area of study and are added without further application of the present algorithm. The final matrix is

$$
\begin{array}{c}
 & 6 & 8 & 7 \\
\begin{array}{c} 6 \\ 8 \\ 7 \end{array}
\left[
\begin{array}{ccc}
0.08999879 & 0.01219496 & 0.03364765 \\
 & 0.06023255 & 0.02708638 \\
 & & 0.07483526
\end{array}
\right]
\end{array}
\qquad \text{bus list } 6, 8, 7
$$

The matrix elements agree exactly with the matrix elements of the example used in the original algorithm.

LARGE EQUIVALENTS

The restriction that a portion of a system to be reduced to an equivalent be of such size that the Z-matrix of the portion to be reduced can be retained in the computer memory no longer applies. The data are reordered by the line ordering routine for the building algorithm. During this reordering the number of lines connected to each bus is counted and recorded. The matrix formation begins, and as each line is processed the line count is corrected to the number of lines remaining to be connected to each bus. When the count of lines connected to any bus is reduced to zero, the row and column corresponding are deleted unless the bus is in the list of buses to be retained. When all lines have been processed the matrix that remains will be the small matrix Z_s discussed on p. 59.

It is therefore possible to process an extremely large system and reduce it to a mesh equivalent of just a few buses. In the matrix formation the matrix grows to a certain size (depending on the degree of interconnection) and then remains rather stationary in size, since buses will be completed at about the same rate that new buses are being added to the system.

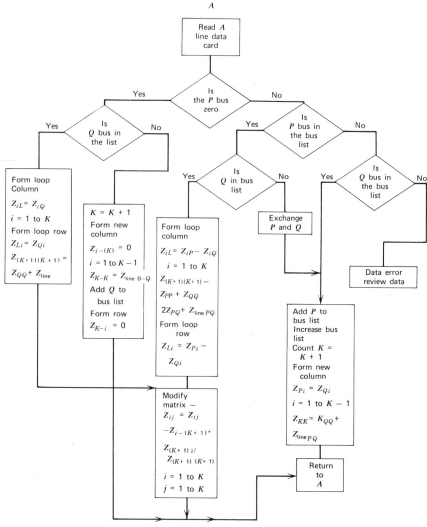

Fig. 4.12

SUMMARY

The technique for extending the Z-bus matrix for short-circuit work to extremely large systems has been described. Automatic reordering of line data into a feasible order for processing by the building algorithm and the reduction of large systems to an equivalent mesh has been detailed.

Unordered line list	System bus list	Reordered line list
(ULL)	(SBL)	(RLL)

Bus numbers		Impedances	
First	Second	Z_+	Z_0
P-bus	Q-bus	Z_1	Z_0

Fig. 4.13. Reordering of line data into feasible order for matrix building.

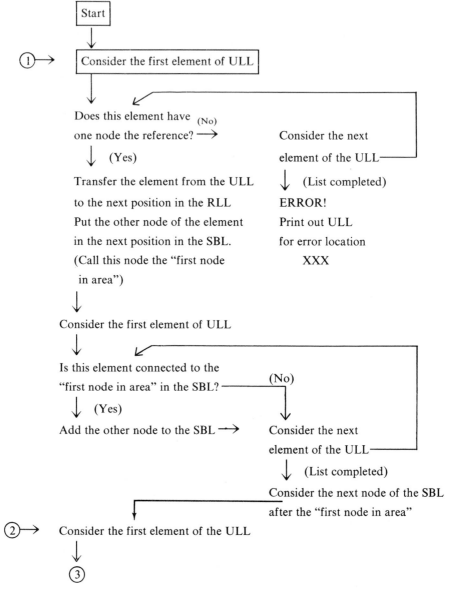

Fig. 4.14. Computer flow chart for line sorting.

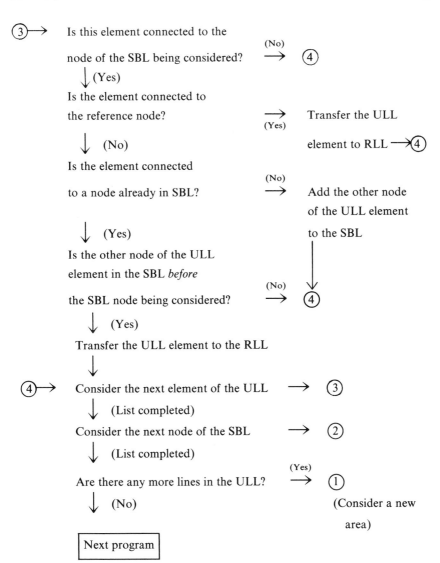

Fig. 4.14. (*Continued*)

References

1. **H. E. Brown, C. E. Person, L. K. Kirchmayer, and G. W. Stagg,** Digital calculation of 3-phase short circuits by matrix method, *Trans. AIEE PA&S*, Vol. 79, (1960), pp. 1277–1282.

2. **R. Baumann,** Some new aspects on load flow calculations, Part I Impedance matrix generation controlled by network topology, *Trans. IEEE PA&S*, Vol. 85, (1966), p. 91.

3. **H. E. Brown and C. E. Person,** Short circuit studies of large systems by the impedance matrix method, *IEEE, PICA Conf. Proc. (1967)*, Pittsburgh, Pa., pp. 335–342.

4. **A. H. El-Abiad,** Digital calculations of line-to-ground short circuits by matrix method, *Trans. AIEE*, Vol. 79, Part III, (1960), p. 323.

Single-Phase
Short Circuits

A three-phase system under single-phase-to-ground short-circuit conditions is an unbalanced system. The method of symmetrical components, which was first described in 1918 by Fortescue [1], is generally used in the solution of unbalanced network problems. For a complete treatment of the material, the reader is referred to textbooks that discuss symmetrical components in some detail, for, example Refs. 2 and 3.

THE MATRIX METHOD

In the three-phase short-circuit analysis of a network, the behavior of the balanced system is computed by means of the Z-bus matrix of only the positive sequence network. This was discussed in Chapter 3. For unbalanced short-circuit conditions (single-line-to-ground, line-to-line, or double-line-to-ground) the three sequence networks (positive, negative, and zero) are connected as required to satisfy the constraints imposed by the unbalanced system condition. In this chapter only the single-line-to-ground problem is discussed. However, the method of computation of the other unbalanced short-circuit conditions follows in similar fashion using the elements available in the matrices.

Generally in power system analysis, the negative sequence network is considered to be identical to the positive sequence network with the exception that there are no sources in the negative sequence network. The Z-bus matrix for the negative sequence is therefore never computed for the obvious reason that computer time and memory are conserved by eliminating this matrix.

The matrices contain impedance equivalents of the actual network. In single-phase-to-ground calculations the three networks are connected to-

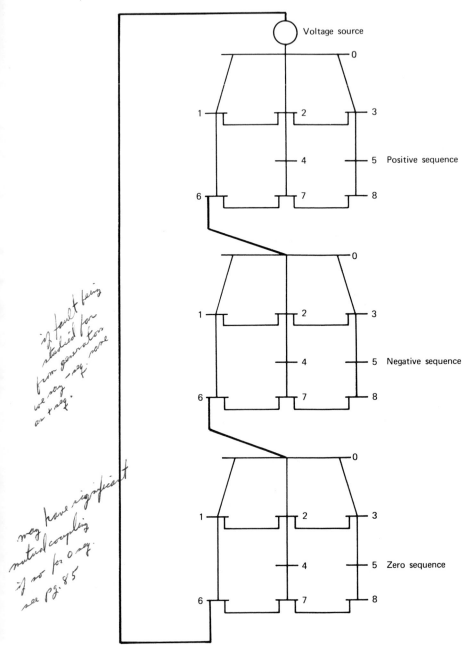

Fig. 5.1. Single-phase-to-ground connection of the three sequence networks.

gether in series as shown for the 8-bus sample network of Ref. 4 as shown in Fig. 5.1 (short circuit on bus 6).

The Z-matrix of the positive sequence network is formed by the building algorithm of Chapter 3, and the axis discarding technique that is described in Chapter 4 is utilized if necessary. After the positive sequence matrix has been formed, it remains in high-speed core storage throughout the remainder of the short circuit analysis. The building algorithm is then used to compute the Z-matrix of the zero sequence network. The information contained in these two matrices enable the complete short-circuit analysis to be made, using only simple arithmetic operations.

For example, the equivalent impedance between bus 6 and the source bus in the positive sequence network is equal to the element Z_{6-6}^{+} in the positive sequence matrix. The element Z_{6-6}^{0} of the zero sequence matrix is the equivalent impedance between bus 6 and the source bus in the zero sequence network. The short-circuit current i_0 which will flow for a short circuit on bus 6 is therefore given by $i_0 = E/(2Z_{6-6}^{+} + Z_{6-6}^{0})$. If $E = 1.0$ pu,

$$i_0 = \frac{1}{2Z_{6-6}^{+} + Z_{6-6}^{0}} \tag{5.1}$$

The total fault current is

$$I = 3i_0 \tag{5.2}$$

The current flowing in any branch during this short-circuit condition can be computed using elements of the column that is subjected to the short-circuit condition. In the example being considered the column corresponding to bus 6 is used.

The flow in line 3-5 in the positive sequence network is obtained as follows:

$$i_{3-5}^{+} = \frac{E_3^{+} - E_5^{+}}{Z_{\text{line 3-5}}^{+}} \tag{5.3}$$

From the relationship $ZI = E$, remembering that all elements in the current vector I are zero except i_6, which is equal to i_0 from equation 5.1, the voltages E_3 and E_5 are obtained from equation 5.4.

$$
\begin{bmatrix}
Z_{1\text{-}6} \\
Z_{2\text{-}6} \\
Z_{3\text{-}6} \\
Z_{4\text{-}6} \\
Z_{5\text{-}6} \\
Z_{6\text{-}6} \\
Z_{7\text{-}6} \\
Z_{8\text{-}6}
\end{bmatrix}
\begin{bmatrix}
\\
\\
\\
\\
\\
i_6 = i_0 \\
\\
\\
\end{bmatrix}
=
\begin{bmatrix}
E_1 \\
E_2 \\
E_3 \\
E_4 \\
E_5 \\
E_6 \\
E_7 \\
E_8
\end{bmatrix}
\tag{5.4}
$$

Note that the difference in voltage between buses 3 and 5 is the same magnitude if the reference bus is at zero voltage and bus 6 is at 1.0 voltage or $E_0 = 1.0$ and $E_6 = 0$, only the sign is changed.

The voltages obtained from equation 5.4 are

$$E_3^+ = i_0 Z_{3\text{-}6}^+ \quad \text{and} \quad E_5^+ = i_0 Z_{5\text{-}6}^+,$$

but the order of these voltages must be reversed in equation 5.3 to account for the change in sign as indicated. This gives

$$i_{3\text{-}5}^+ = \frac{i_0(Z_{5\text{-}6}^+ - Z_{3\text{-}6}^+)}{Z_{\text{line }3\text{-}5}^+}$$

in which the substitution of the value of i_0 from equation 5.1 gives equation 5.5.

$$i_{3-5}^+ = \frac{1}{2Z_{6-6}^+ + Z_{6-6}^0} \frac{(Z_{5-6}^+ - Z_{3-6}^+)}{Z_{\text{line }3-5}^+} \tag{5.5}$$

The current flowing in the line in the zero sequence network is obtained by using the zero sequence elements:

$$i_{3-5}^0 = \frac{1}{2Z_{6-6}^+ + Z_{6-6}^0} \frac{(Z_{5-6}^0 - Z_{3-6}^0)}{Z_{\text{line }3-5}^0} \tag{5.6}$$

The total current in the line is:

$$i_{3-5} = i_{3-5}^+ + i_{3-5}^- + i_{3-5}^0 = 2i_{3-5}^+ + i_{3-5}^0$$

SYSTEM CHANGES

The system may be changed corresponding to line additions and removals as discussed in Chapter 3, with the exception that now two matrices must be modified to reflect the new system condition.

AN ADDITIONAL COMPLICATION

Transmission lines that share the same right-of-way will be magnetically coupled in the zero sequence network. Account of this coupling must be taken in obtaining the zero sequence matrix. The method of adjusting the matrix elements to account for this mutual coupling has gone through a series of developments.

1. The zero sequence matrix was formed without mutual couplings and then modified for all couplings simultaneously [5]. This method is similar to applying a Kron reduction to n axes simultaneously when forming the Z-matrix.
2. A subsequent method formed the matrix and later compensated for the couplings one at a time by a two-step operation [6]. This is similar to a Kron reduction eliminating one axis at a time from the complete matrix.
3. The latest method used was suggested by Reitan [7]. In this method the matrix is adjusted for the coupling at the time the mutually coupled line is added to the network. The earlier methods are of historical interest. The only method to be described here is the last method, which is superior to the other two methods.

MUTUAL COUPLING COMPENSATION

The building of the zero sequence matrix is complicated by the existence of magnetic couplings in the zero sequence network. The formation of the matrix follows the line-by-line addition as described earlier but with one essential difference. The building algorithm is modified, for mutually coupled lines, to include the effect of the coupling at the time the line is added to the network. Addition of lines not mutually coupled and the addition of the first line on a right-of-way follow the matrix building algorithm previously described.

THE MODIFIED LOGIC

As each line is selected from the ordered line list for processing, it is checked against the list of coupled lines. If it is not found in the coupling list, the line is processed by the original building algorithm according to the type of line, radial, and loop closing. If the line is found to be a line in the mutual coupling list, a further check is made to find if this is the first line of the mutually coupled set to be added to the system. The original building algorithm is used if it is the first line on a right-of-way, since a line can not be coupled to lines that at this point do not exist and have not been incorporated into the network. However, if the line to be added is coupled to lines in the system the modification of the matrix must take into account the mutual coupling. Two routines must be used depending on whether the line is a radial line from an existing bus to a new bus or if it is a loop closing line. These logical decisions are detailed on Fig. 5.2.

MODIFIED RADIAL LINE ALGORITHM

A line has been selected from the data list and has been found to be mutually coupled to a line (or lines) that have already been included in the network.

The line to be added is a line PQ which is a radial line from bus P, which is already included in the system, to a new bus Q. The line is mutually coupled to one or more line RS. Here RS is a running index (see Fig. 5.3).

The relationship $I = YE$ that applies to the new line PQ and the several lines to which it is coupled is given by

$$
\begin{bmatrix} i_{PQ} \\ i_{RS} \end{bmatrix} = \begin{bmatrix} Y_{PQ} & Y_{PQ\text{-}RS} \\ Y_{RS\text{-}PQ} & Y_{RS\text{-}RS} \end{bmatrix} \begin{bmatrix} v_{PQ} \\ v_{RS} \end{bmatrix} \qquad (5.7)
$$

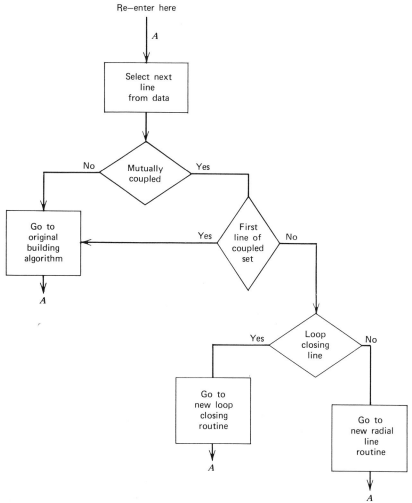

Fig. 5.2. Logic of the modified building algorithm.

From equation 5.7

$$i_{PQ} = Y_{PQ}v_{PQ} + [Y_{PQ\text{-}RS}][v_{RS}] \tag{5.8}$$

Note $[Y_{PQ\text{-}RS}]$ is a matrix row and v_{RS} is a column vector.

For the case $K \neq Q$ Injection of unit current into any bus K of the system will produce $i_{PQ} = 0$, since line PQ is a radial line from P to Q and is not a part of a mesh. Equation 5.8 becomes

$$Y_{PQ}v_{PQ} + [Y_{PQ\text{-}RS}][v_{RS}] = 0$$

which can be written

$$v_{PQ} = \frac{-[Y_{PQ\text{-}RS}][v_{RS}]}{Y_{PQ}} \tag{5.9}$$

since the voltage across a line is equal to the difference of voltages of the buses at the ends of the line,

$$v_{PQ} = E_P - E_Q \tag{5.10}$$

From the matrix equation $ZI = E$ where only column K of the Z-matrix is explicity shown and the current vector has the single nonzero element $i_K = 1.0$, the voltages of all the system buses are determined for unit current injection into bus K (see equation 5.11).

$$
\begin{bmatrix}
\cdots & Z_{1K} & \cdots \\
\cdots & Z_{2K} & \cdots \\
\cdots & Z_{3K} & \cdots \\
\cdots & \cdot & \cdots \\
\cdots & \cdot & \cdots \\
\cdots & \cdot & \cdots \\
\cdots & Z_{PK} & \cdots \\
\cdots & \cdot & \cdots \\
\cdots & \cdot & \cdots \\
\cdots & \cdot & \cdots \\
\cdots & Z_{QK} & \cdots \\
\cdots & \cdot & \cdots \\
\cdots & \cdot & \cdots \\
\cdots & \cdot & \cdots
\end{bmatrix}
\begin{bmatrix}
0 \\ 0 \\ 0 \\ \cdot \\ \cdot \\ \cdot \\ \cdot \\ \cdot \\ \cdot \\ i_K = 1.0 \\ \cdot \\ \cdot \\ \cdot \\ \cdot
\end{bmatrix}
=
\begin{bmatrix}
E_1 \\ E_2 \\ E_3 \\ \cdot \\ \cdot \\ \cdot \\ E_P \\ \cdot \\ \cdot \\ E_Q \\ \cdot \\ \cdot
\end{bmatrix}
\tag{5.11}
$$

Substitution of equation 5.10 into 5.9

$$E_P - E_Q = \frac{-[Y_{PQ\text{-}RS}][E_R - E_S]}{Y_{PQ}} \tag{5.12}$$

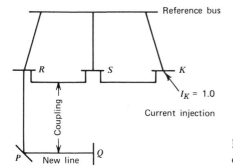

Fig. 5.3. Network with new radially connected mutually coupled line.

From equation 5.11 the values of the bus voltages

$$E_P = Z_{PK}$$

$$E_Q = Z_{QK}$$

$$E_R = Z_{RK}$$

$$E_S = Z_{SK}$$

are obtained. Substitution of these values into 5.12

$$Z_{PK} - Z_{QK} = \frac{-[Y_{PQ\text{-}RS}][Z_{RK} - Z_{SK}]}{Y_{PQ}} \tag{5.13}$$

$$Z_{QK} = Z_{PK} + \frac{[Y_{PQ\text{-}RS}][Z_{RK} - Z_{SK}]}{Y_{PQ}} \tag{5.14}$$

As K takes on all values $K = 1, 2, 3 \ldots n$ but $\neq Q$ the elements of the new column of the matrix corresponding to bus Q are obtained. The RS take on all values of lines of the system coupled to the new line PQ. It remains to obtain the diagonal element of the new matrix axis corresponding to Q.

For the case $K = Q$ The diagonal element is determined by injecting unit current into bus Q as shown in Fig. 5.4.

In equation 5.8 i_{PQ} is now equal to -1.0, since $i_Q = i_{QP} = -i_{PQ} = 1.0$ as can be seen in Fig. 5.4.

$$Y_{PQ} v_{PQ} + [Y_{PQ\text{-}RS}][v_{RS}] = -1.0 \tag{5.15}$$

$$v_{PQ} = -\frac{1.0 + [Y_{PQ\text{-}RS}][v_{RS}]}{Y_{PQ}} \tag{5.16}$$

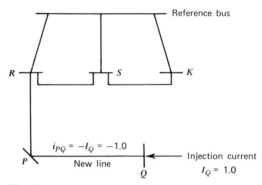

Fig. 5.4

Again $v_{PQ} = E_P - E_Q$ and $v_{RS} = E_R - E_S$. Since $I_Q = 1.0$ the values

$$\left.\begin{array}{c} E_R = Z_{RQ} \\ E_S = Z_{SQ} \\ \cdot \quad \cdot \\ \cdot \quad \cdot \\ \cdot \quad \cdot \end{array}\right\}$$

were computed by equation 5.14,

$$v_{PQ} = E_P - E_Q = Z_{PQ} - Z_{QQ}$$

and

$$v_{RS} = E_R - E_S = Z_{RQ} - Z_{SQ}$$

equation 5.16 then can be written

$$Z_{PQ} - Z_{QQ} = -\frac{1.0 + [Y_{PQ\text{-}RS}][Z_{RQ} - Z_{SQ}]}{Y_{PQ}}$$

and rearranged as shown in equation 5.17.

$$Z_{QQ} = Z_{PQ} + \frac{1.0 + [Y_{PQ\text{-}RS}][Z_{RQ} - Z_{SQ}]}{Y_{PQ}} \tag{5.17}$$

The off diagonal elements for the new column for the matrix are computed by equation 5.14 and are used in 5.17 to determine the new diagonal element.

ADDITION OF A LOOP CLOSING <u>MUTUALLY COUPLED LINE</u>

Addition of a new line PQ between two buses P and Q, which have already been included in the system, requires a modified algorithm if the new line is mutually coupled to lines of the system that has been assembled. The new line PQ is coupled to the lines RS, where RS is a running index of the lines included in the system.

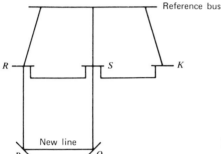

Fig. 5.5. Network with a new mutually coupled loop closing line.

Addition of the new line does not provide a new bus for the network and the matrix. New constraints are supplied by the loop that is created when the line PQ is added to the system. A loop axis is therefore added to the matrix as was done in Chapter 3 on p. 30. The algorithm developed in Chapter 3 is modified to include the effect of the mutual coupling.

Unit current is injected into bus K, $K = 1, 2, 3, \ldots n$, but $K \neq Q$. A voltage source e_{loop} is introduced in the new line PQ and adjusted so that $i_{PQ} = 0$ when $I_K = 1.0$ as shown in Fig. 5.6.

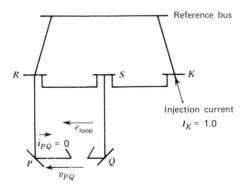

Fig. 5.6

The voltage v_{PQ} is the voltage across the line PQ.

$$e_{\text{loop}} + v_{PQ} = E_P - E_Q$$

or

$$e_{\text{loop}} = E_P - E_Q - v_{PQ} \tag{5.18}$$

Since unit current is being injected into bus K

$$E_P = Z_{PK}$$
$$E_Q = Z_{QK}$$
$$E_R = Z_{RK}$$
$$\tag{5.19}$$
$$e_{\text{loop}} = Z_{\text{loop-}K}$$

from consideration of Fig. 5.6 and the understanding of the meaning of the Z-matrix.

Returning to the matrix equation 5.7 and obtaining the expression for the current

$$i_{PQ} = Y_{PQ} v_{PQ} + [Y_{PQ\text{-}RS}][v_{RS}] \tag{5.20}$$

The voltage source e_{loop} was so adjusted that $i_{PQ} = 0$. Equation 5.20 can therefore be written

$$v_{PQ} = -\frac{[Y_{PQ\text{-}RS}][v_{RS}]}{Y_{PQ}} \tag{5.21}$$

Substitution of this value into equation 5.18 gives

$$e_{\text{loop}} = E_P - E_Q + \frac{[Y_{PQ\text{-}RS}][v_{RS}]}{Y_{PQ}} \tag{5.22}$$

since $v_{RS} = E_R - E_S$.

$$e_{\text{loop}} = E_P - E_Q + \frac{[Y_{PQ\text{-}RS}][E_R - E_S]}{Y_{PQ}} \tag{5.23}$$

Replacing the voltages in equation 5.23 by the values from equation 5.9. (the current injection $I_K = 1.0$) gives

$$Z_{\text{loop-}K} = Z_{PK} - Z_{QK} + \frac{[Y_{PQ\text{-}RS}][Z_{RK} - Z_{SK}]}{Y_{PQ}} \qquad (5.24)$$

As K takes on values $1, 2, 3, \ldots$ the loop row and column elements, exclusive of the diagonal element, are obtained.

The diagonal element   To obtain the diagonal element the voltage source e_{loop} is adjusted to cause unit current i_{QP} to flow around the loop (see Fig. 5.7). Equation 5.16 applies to this case

$$v_{PQ} = - \frac{1.0 + [Y_{PQ\text{-}RS}][v_{RS}]}{Y_{PQ}}$$

When this equation is substituted into equation 5.18

$$e_{\text{loop}} = E_P - E_Q + \frac{1.0 + [Y_{PQ\text{-}RS}][v_{RS}]}{Y_{PQ}} \qquad (5.25)$$

Unit current is forced to flow in the loop by adjusting e_{loop} to the required value. In this case $e_{\text{loop}} = Z_{\text{loop-loop}}$. That is, the voltage required to drive a loop current of 1.0 around the loop is by definition equal to the driving point impedance of the loop.

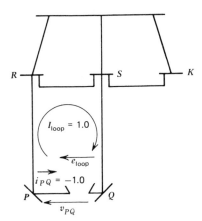

Fig. 5.7.   The loop constraint.

Replacing $E_P, E_Q, v_{RS} = E_R - E_S$ by the appropriate terms, $Z_{P\text{-loop}}$, $Z_{Q\text{-loop}}$, $Z_{R\text{-loop}}$, $Z_{S\text{-loop}}$, equation 5.25 becomes

$$Z_{\text{loop-loop}} = Z_{P\text{-loop}} - Z_{Q\text{-loop}} + \frac{1.0 + [Y_{PQ\text{-}RS}][Z_{R\text{-loop}} - Z_{S\text{-loop}}]}{Y_{PQ}} \quad (5.26)$$

Note: All elements $Z_{P\text{-loop}}, Z_{Q\text{-loop}} \ldots$, were previously determined by use of equation 5.24.

Example The following is a numerical example of the modified matrix building algorithm. All logical decisions that must be made according to Fig. 5.2 are shown as well as the numerical work required to build the matrix. The network that is considered is the network given in the El-Abiad paper [5]. The diagram of the network is given in the Fig. 5.8 and the following data:

Line Data

Line	Reactance
0-1	0.1
0-2	0.5
1-2	0.4
0-3	0.5
1-4	0.2
4-3	0.3

Mutual Coupling Impedance Data

Lines		Mutual Impedance
1-2	1-4	0.1
1-2	4-3	0.2

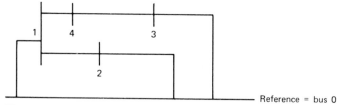

Fig. 5.8. Network to be assembled by the building algorithm.

The line data have been ordered in a feasible sequence for processing. Zero is stored in the memory area allocated to the matrix and the bus list of the system that will be assembled.

1. The first line is selected for processing, line 0-1. The line is identified as a line from the reference bus 0. Bus 1 is not in the bus list and the line is therefore a line from the reference to a new bus. This is a degenerate case, since the matrix and bus list are zero. The new axis corresponding to the new bus 1 is added to the matrix. The impedance of the line is placed at the diagonal and bus 1 is added to the bus list.

$$\text{matrix} = 1 \begin{array}{c} 1 \\ [.1] \end{array} \qquad \text{bus list } 1$$

2. The line 0-2 is also identified to be a line from the reference. The matrix is augmented by a row and column of zeros. The impedance of the line is placed at the diagonal. Bus number 2 is added to the bus list.

$$\text{matrix} \begin{array}{c} \\ 1 \\ 2 \end{array} \begin{bmatrix} 0.1 & 0 \\ 0 & 0.5 \end{bmatrix} \qquad \text{bus list } 1,2$$

3. The line 1-2 is identified to be a loop closing line, since both buses describing the line are already in the bus list. A search of the lines involved in mutual couplings show that the line is mutually coupled but neither of the lines to which it is coupled have been added to the system. The original loop closing algorithm (without mutual) is used (see Fig. 5.2).

$$\text{matrix} \begin{array}{c} \\ 1 \\ 2 \\ \text{loop} \end{array} \begin{bmatrix} 0.1 & 0 & 0.1 \\ 0 & 0.5 & -0.5 \\ 0.1 & -0.5 & 1.0 \end{bmatrix} \qquad \text{bus list } 1, 2$$

The matrix is reduced by a Kron reduction.

$$\text{matrix} \begin{array}{c} \\ 1 \\ 2 \end{array} \begin{bmatrix} 0.09 & 0.05 \\ 0.05 & 0.25 \end{bmatrix} \qquad \text{bus list } 1, 2$$

4. Addition of the line 0-3 follows the example of 2 above.

$$
\begin{array}{c}
\begin{array}{ccc} \quad 1 \quad & \quad 2 \quad & \quad 3 \end{array}\\
\text{matrix}\; \begin{array}{c} 1 \\ 2 \\ 3 \end{array}
\left[
\begin{array}{ccc}
0.09 & 0.05 & 0 \\
0.05 & 0.25 & 0 \\
0 & 0 & 0.5
\end{array}
\right]
\end{array}
\qquad \text{bus list 1, 2, 3}
$$

5. Comparison of the line 1-4 with the mutual coupling data shows that this new line is mutually coupled to the line 1-2 which is already in the system. Examination of the bus list shows that line 1-4 is a radial line from an existing bus (1) to a new bus (4). The routine that is required is the mutually coupled radial line routine. Form the coupling matrix where the diagonal elements are the self-impedances of the conductors and the off diagonals are the mutual impedances between lines.

$$
X_{\text{coupling}} = \begin{array}{c}
\begin{array}{cc} \quad 1\text{-}4 \quad & \quad 1\text{-}2 \end{array}\\
\begin{array}{c} 1\text{-}4 \\ 1\text{-}2 \end{array}
\left[
\begin{array}{cc}
0.2 & 0.1 \\
0.1 & 0.4
\end{array}
\right]
\end{array}
$$

Inversion gives

$$
X_{\text{coupling}}^{-1} = Y_{\text{coupling}} = \begin{array}{c}
\begin{array}{cc} \qquad 1\text{-}4 \qquad & \qquad 1\text{-}2 \end{array}\\
\begin{array}{c} 1\text{-}4 \\ 1\text{-}2 \end{array}
\left[
\begin{array}{cc}
5.7143 & -1.4286 \\
-1.4286 & 2.8571
\end{array}
\right]
\end{array}
$$

For a radial line apply equation 5.14

$$
Z_{Qi} = Z_{Pi} + \frac{[Y_{PQ\text{-}RS}][Z_{Ri} - Z_{Si}]}{Y_{PQ}}
$$

It must be pointed out that at this point the computation for this line is a very degenerate case, since the line $PQ(1\text{-}4)$ is coupled to a single line $RS(1\text{-}2)$.

$$
Y_{PQ} = Y_{14} = 5.7143
$$

From the coupling Y-matrix

$$
[Y_{PQ\text{-}RS}] = [Y_{14\text{-}12}] = [-1.4286]
$$

The first element in the new column for bus 4 is

$$Z_{41} = Z_{11} + \frac{[Y_{14\text{-}12}][Z_{11} - Z_{21}]}{Y_{14}}$$

The elements Z_{11}, Z_{21} are obtained from the Z-matrix that is being built.

$$Z_{41} = 0.09 + \frac{[-1.4286][0.09 - 0.05]}{5.7143} = 0.09 - 0.01 = 0.08$$

$$Z_{42} = Z_{12} + \frac{[-1.4286][Z_{12} - Z_{22}]}{5.7143}$$

$$Z_{42} = 0.05 + \frac{[-1.4286][0.05 - 0.25]}{5.7143} = 0.05 + 0.05 = 0.10$$

$$Z_{43} = Z_{13} + \frac{[-1.4286][Z_{13} - Z_{23}]}{5.7143} = + \frac{[-1.4286][0 - 0]}{5.7143} = 0$$

The diagonal element is computed by equation 5.17

$$Z_{QQ} = Z_{PQ} + \frac{1.0 + [Y_{PQ\text{-}RS}][Z_{RQ} - Z_{SQ}]}{Y_{PQ}}$$

The values Z_{RQ}, Z_{PQ}, and Z_{SQ} are the values that have just been computed.

$$Z_{44} = Z_{14} + \frac{1.0 + [-1.4286][Z_{14} - Z_{24}]}{5.7143}$$

$$Z_{44} = 0.08 + \frac{1.0 + [-1.4286][0.08 - 0.10]}{5.7143}$$

$$Z_{44} = 0.08 + 0.18 = 0.26$$

	1	2	3	4
1	0.09	0.05	0	0.08
2	0.05	0.25	0	0.10
3	0	0	0.5	0
4	0.08	0.10	0	0.26

matrix   bus list 1, 2, 3, 4

Mutually coupled lines included

$$1\text{-}4$$
$$1\text{-}2$$

6. The line 4-3 is identified to be a line that is mutually coupled to the line 1-2. The coupling matrix describing the right-of-way that includes the line 1-2 is augmented to include the line 4-3.

$$X_{\text{coupling}} = \begin{array}{c} 4\text{-}3 \\ 1\text{-}4 \\ 1\text{-}2 \end{array} \begin{array}{ccc} 4\text{-}3 & 1\text{-}4 & 1\text{-}2 \\ \left[\begin{array}{ccc} 0.3 & 0 & 0.2 \\ 0 & 0.2 & 0.1 \\ 0.2 & 0.1 & 0.4 \end{array} \right] \end{array}$$

Inversion of X_{coupling} gives

$$X_{\text{coupling}}^{-1} = Y_{\text{coupling}} = \begin{array}{c} 4\text{-}3 \\ 1\text{-}4 \\ 1\text{-}2 \end{array} \begin{array}{ccc} 4\text{-}3 & 1\text{-}4 & 1\text{-}2 \\ \left[\begin{array}{ccc} 5.38461 & 1.53846 & -3.07692 \\ 1.53846 & 6.15384 & -2.30769 \\ -3.07692 & -2.30769 & 4.61538 \end{array} \right] \end{array}$$

The element $Y_{43} = 5.38461$ and is the diagonal element of the Y_{coupling} matrix corresponding to the new line 4-3.

Also required are the off diagonal elements of the coupling matrix corresponding to the new line 4-3.

$$[Y_{PQ\text{-}RS}] = [Y_{43\text{-}14} \, Y_{43\text{-}12}]$$

$$= [1.53846 - 3.07692]$$

Examination of the bus list identifies that the new line is a loop closing line. A loop axis is added to the matrix by using equation 5.24 for the off diagonal elements and equation 5.26 for the diagonal element.

$$Z_{\text{loop-}K} = Z_{PK} - Z_{QK} + \frac{[Y_{PQ\text{-}RS}][Z_{RK} - Z_{SK}]}{Y_{PQ}} \qquad (5.24)$$

$$Z_{\text{loop-1}} = Z_{41} - Z_{31} + \frac{[Y_{43\text{-}14}\, Y_{43\text{-}12}] \begin{bmatrix} Z_{11} - Z_{41} \\ Z_{11} - Z_{21} \end{bmatrix}}{Y_{43}}$$

$$= 0.08 - 0 + \frac{[1.53846 - 3.07692] \begin{bmatrix} 0.09 - 0.08 \\ 0.09 - 0.05 \end{bmatrix}}{5.38461}$$

$$= 0.08 + \frac{[1.53846 - 3.07692] \begin{bmatrix} 0.01 \\ 0.04 \end{bmatrix}}{5.38461} = 0.08 - 0.02 = 0.06$$

$$Z_{\text{loop-2}} = Z_{42} - Z_{32} + \frac{[1.53846 - 3.07692] \begin{bmatrix} Z_{12} - Z_{42} \\ Z_{12} - Z_{22} \end{bmatrix}}{5.38461}$$

$$= 0.10 - 0 + \frac{[1.53846 - 3.07692] \begin{bmatrix} 0.05 - 0.10 \\ 0.05 - 0.25 \end{bmatrix}}{5.38461}$$

$$= 0.10 + \frac{[1.53846 - 3.07692] \begin{bmatrix} -0.05 \\ -0.20 \end{bmatrix}}{5.38461}$$

$$Z_{\text{loop-2}} = 0.10 + 0.10 = 0.20$$

$$Z_{\text{loop-3}} = Z_{43} - Z_{33} + \frac{[1.53846 - 3.07692]\begin{bmatrix} Z_{13} - Z_{43} \\ Z_{13} - Z_{23} \end{bmatrix}}{5.38461}$$

$$= 0 - 0.5 + \frac{[1.53846 - 3.07692]\begin{bmatrix} 0 - 0 \\ 0 - 0 \end{bmatrix}}{5.38461}$$

$$= -0.5$$

$$Z_{\text{loop-4}} = Z_{44} - Z_{34} + \frac{[1.53846 - 3.07692]\begin{bmatrix} Z_{14} - Z_{44} \\ Z_{14} - Z_{24} \end{bmatrix}}{5.38461}$$

$$= 0.26 - 0 + \frac{[1.53846 - 3.07692]\begin{bmatrix} 0.08 - 0.26 \\ 0.08 - 0.10 \end{bmatrix}}{5.38461}$$

$$= 0.26 - 0.04 = 0.22$$

The value of $Z_{\text{loop-loop}}$ is computed by means of equation 5.26 using the values that have just been computed.

$$Z_{\text{loop-loop}} = Z_{4\text{-loop}} - Z_{3\text{-loop}} + \frac{1.0 + [Y_{43\text{-}14}\, Y_{43\text{-}12}]\begin{bmatrix} Z_{1\text{-loop}} - Z_{4\text{-loop}} \\ Z_{1\text{-loop}} - Z_{2\text{-loop}} \end{bmatrix}}{Y_{43}}$$

$$Z_{\text{loop-loop}} = 0.22 + 0.5 + \frac{1.0 + [1.53846 - 3.07692]\begin{bmatrix} 0.06 - 0.22 \\ 0.06 - 0.20 \end{bmatrix}}{5.38461}$$

$$= 0.72 + \frac{1.0 + 0.18461}{5.38461} = 0.94$$

The augmented matrix is therefore

	1	2	3	4	loop
1	0.09	0.05	0	0.08	0.06
2	0.05	0.25	0	0.10	0.20
3	0	0	0.5	0	−0.50
4	0.08	0.10	0	0.26	0.22
loop	0.06	0.20	−0.50	0.22	0.94

which is reduced by the Kron reduction

	1	2	3	4
1	0.08617	0.03723	0.03191	0.06596
2	0.03723	0.20745	0.10638	0.05319
3	0.03191	0.10638	0.23405	0.11702
4	0.06596	0.05319	0.11702	0.20851

bus list 1, 2, 3, 4

Lines mutually coupled included

1-2   1-4
1-2   4-3

OPENING A MUTUALLY COUPLED LINE

It was thought for some time that it was impossible to open a line by the method of adding a line with the negative of original line data when a line was mutually coupled (see p. 89 of Ref. 8). The correct procedure in this case was first described by Reitan and Kruemple [7].

The procedure is to add a line to the right-of-way with the negative of the original line impedance but with mutual couplings the same as the original line.

For example to remove the line 1-4 of the previous example add a line (1-4)'to the coupling matrix.

$$
X_{\text{coupling}} = \begin{array}{c} \\ 1\text{-}2 \\ 1\text{-}4 \\ 4\text{-}3 \\ (1\text{-}4)' \end{array}
\begin{array}{cccc} 1\text{-}2 & 1\text{-}4 \ 4\text{-}3 & (1\text{-}4)' \\ \left[\begin{array}{cccc} 0.4 & 0.1 & 0.2 & 0.1 \\ 0.1 & 0.2 & 0 & 0 \\ 0.2 & 0 & 0.3 & 0 \\ 0.1 & 0 & 0 & -0.2 \end{array}\right] \end{array}
$$

The couplings of line (1-4)' are the same as the couplings of line 1-4. The diagonal element of (1-4)' is the negative of diagonal element of the original line 1-4.

$$
X_{\text{coupling}}^{-1} = Y_{\text{coupling}} = \begin{array}{c} \\ 1\text{-}2 \\ 1\text{-}4 \\ 4\text{-}3 \\ (1\text{-}4)' \end{array}
\begin{array}{cccc} 1\text{-}2 & 1\text{-}4 & 4\text{-}3 & (1\text{-}4)' \\ \left[\begin{array}{cccc} 3.75 & -1.875 & -2.5 & 1.875 \\ -1.875 & 5.9375 & 1.25 & -0.9375 \\ -2.5 & 1.25 & 5.00 & -1.25 \\ 1.875 & -0.9375 & -1.25 & -4.0625 \end{array}\right] \end{array}
$$

From Y_{coupling} the row matrix is obtained

$$
[Y_{PQ\text{-}RS}] = (1\text{-}4)' \begin{array}{ccc} 1\text{-}2 & 1\text{-}4 & 4\text{-}3 \\ [1.875 & -0.9375 & -1.25] \end{array}
$$

and the element $Y_{PQ} = (1\text{-}4)' \begin{array}{c} (1\text{-}4)' \\ (-4.0625) \end{array}$. Removal of the line then follows the loop closing line of the previous example. The unreduced matrix is given for reference. The verification is left as an exercise for the reader.

	1	2	3	4	loop
1	0.08617	0.03722	0.03191	0.06596	0.01276
2	0.03722	0.20747	0.10639	0.05320	0.04254
3	0.03191	0.10639	0.23404	0.11703	−0.10639
4	0.06596	0.05320	0.11703	0.20851	−0.15319
loop	0.01276	0.04254	−0.10639	−0.15319	−0.04255

The Kron reduction eliminates the loop axis and the matrix that results is

	1	2	3	4
1	0.09	0.05	0	0.02
2	0.05	0.25	0	−0.1
3	0	0	0.5	0.5
4	0.02	−0.1	0.5	0.76

Exercise   Show that the line data

0-1	0.1
0-2	0.5
1-2	0.4
0-3	0.5
4-3	0.3

with mutual coupling between line 1-2 and line 4-3 of 0.2 gives the same matrix as the system used in the example after the line 1-4 was removed.

Note: In assembly of the system the line 4-3 must be added from 3 to 4 because 3 is in the system and 4 is a new bus. The sign of the mutual coupling between lines 1-2 and 4-3 of 0.2 must be considered carefully.

FLOW IN LINES IN THE ZERO SEQUENCE NETWORK

In single-phase-to-ground studies the flow in a line in the positive sequence network is obtained by a modification of equation 3.5 of chapter 3.

$$i_{PQ}^+ = \frac{Z_{QK}^+ - Z_{PK}^+}{Z_{\text{line } PQ}^+} \frac{1}{2Z_{KK}^+ + Z_{KK}^0} \tag{5.27}$$

The flow in a line in the zero sequence network that is not mutually coupled is computed using a similar formula.

$$i_{PQ}^0 = \frac{Z_{QK}^0 - Z_{PK}^0}{Z_{\text{line } PQ}^0} \frac{1}{2Z_{KK}^+ + Z_{KK}^0} \tag{5.28}$$

When a line is mutually coupled the flow in the zero sequence network is computed by using one row from the matrix equation 5.7.

$$i_{PQ} = [Y_{PQ} \; Y_{PQ\text{-}RS}] \begin{bmatrix} v_{PQ} \\ v_{RS} \end{bmatrix}$$

$$i_{PQ} = Y_{PQ} v_{PQ} + [Y_{PQ\text{-}RS}][v_{RS}] \tag{5.29}$$

since

$$v_{PQ} = E_P - E_Q = i_K(Z_{QK} - Z_{PK})$$

$$v_{RS} = E_R - E_S = i_K(Z_{SK} - Z_{RK})$$

and

$$i_K = \frac{1}{2Z_{KK}^+ + Z_{KK}^0}$$

Equation 5.28 then becomes

$$i_{PQ}^0 = \frac{1}{2Z_{KK}^+ + Z_{KK}^0} [Y_{PQ}(Z_{QK}^0 - Z_{PK}^0) + [Y_{PQ\text{-}RS}][Z_{SK}^0 - Z_{RK}^0]]$$

The total phase current in a line PQ is then obtained by combining the line currents of the three networks.

$$i_{PQ} = 2i_{PQ}^+ + i_{PQ}^0 \tag{5.30}$$

Example   The positive sequence matrix of the network used in the previous numerical example from Ref. 5 is required. It is given here without being computed.

$$
Z^+ = \begin{array}{c} \\ 1 \\ 2 \\ 3 \\ 4 \end{array}
\begin{array}{cccc}
1 & 2 & 3 & 4 \\
\end{array}
\left[
\begin{array}{cccc}
0.03944 & 0.02254 & 0.01970 & 0.02958 \\
0.02254 & 0.09859 & 0.01128 & 0.01690 \\
0.01970 & 0.01128 & 0.10986 & 0.06479 \\
0.02958 & 0.01690 & 0.06479 & 0.09718
\end{array}
\right]
$$

The flow in lines can be reported in amperes per unit or MVA by adjusting the base of the study. If the line impedances were on 100 MVA base the total short circuit per unit current could be computed by equation 5.31 when 1.0-pu voltage is applied to the network.*

$$
3I_0 = \frac{3\,\text{base}}{2Z_{KK}^+ + Z_{KK}^0} = \frac{(3)(1.0)}{2Z_{KK}^+ + Z_{KK}^0}\,\text{pu} \tag{5.31}
$$

For a short circuit on bus 1 we would obtain

$$
3I_0 = \frac{3}{(2)(0.03944) + 0.08617} = 18.18\,\text{pu}
$$

The current flowing through the 3 networks in series would be $I_0 = 6.06\,\text{pu}$.

To obtain the flow in the line 1-2 in the zero sequence network for a short circuit, elements from the coupling admittance matrix (Y_{PQ} and $|Y_{PQ\text{-}RS}|$) are used as well as the elements from the zero sequence bus impedance matrix (Z_{PK}^0, Z_{QK}^0, etc.).

$$
I_{21}^0 = I_0[\, Y_{12}\; Y_{12\text{-}14}\; Y_{12\text{-}43}\,]
\left[
\begin{array}{c}
Z_{11} - Z_{12} \\
Z_{11} - Z_{14} \\
Z_{14} - Z_{13}
\end{array}
\right]
$$

$$
= 6.06[\,4.615 - 2.308 - 3.077\,]
\left[
\begin{array}{c}
0.08617 - 0.03723 \\
0.08617 - 0.06596 \\
0.06596 - 0.03191
\end{array}
\right]
$$

$$
= 6.06(0.2258 - 0.0465 - 0.1048) = 6.06(0.0745)
$$

$$
= 0.451
$$

*A practice that has become rather common among relay engineers is to refer to the total fault current $3I$ and flow in lines as amperes or MVA interchangeably.

To obtain the flow in another line in the mutually coupled group the Z-matrix vector remains the same. The only change is that a different row of the mutual coupling matrix is used. For a fault on bus 1 the flow in the line 4-1 is computed as follows

$$I_{41}^0 = 6.06[\, Y_{14\text{-}12} \; Y_{14} \; Y_{14\text{-}43}] \begin{bmatrix} Z_{11} - Z_{12} \\ Z_{11} - Z_{14} \\ Z_{14} - Z_{13} \end{bmatrix}$$

$$= 6.06[\, -2.308 \quad 6.154 \quad 1.538] \begin{bmatrix} 0.04894 \\ 0.02021 \\ 0.03405 \end{bmatrix}$$

$$= 6.06(-0.1130 + 0.1244 + 0.0523) = 6.06(0.0640) = 0.388$$

The contribution of the only other line connected to bus 1 is from the line 0-1.

The current flow in this line is obtained as follows. The voltage on bus 1 when bus 1 is subjected to unit injection current is $E_1 = Z_{11}^0$. The current injection is $I_0 = 6.06$ pu.

Therefore $E_1' = (I_0) \times (Z_{11}^0) = (6.06)(0.08617) = 0.522$ pu and remembering that $E_0 = 0$ the current flowing in the line 0-1 whose impedance is 0.1 pu is

$$I_{0\text{-}1}^0 = \frac{E_1 - E_0}{Z_{\text{line}\,0\text{-}1}} = \frac{0.522}{0.1} = 5.22$$

The total fault current 6.06 is the sum of the three line flows that have just been obtained.

References

1. **C. L. Fortescue,** Method of symmetrical coordinates applied to the solution of polyphase networks, *Trans. AIEE,* Vol. 37, (1918), p. 1027.

2. **E. Clarke,** *Circuit Analysis of A. C. Power Systems, Symmetrical and Related Components,* Vol. 1, Wiley, 1943.

3. **C. F. Wagner and R. D. Evans,** *Symmetrical Components,* McGraw-Hill, 1933.

4. **H. E. Brown, C. E. Person, L. K. Kirchmayer, and G. W. Stagg,** Digital calculation of three-phase short circuits by matrix method, *Trans. IEEE*, Vol. 79, Part 3, (1960), p. 1277.

5. **A. H. El-Abiad,** Digital calculations of line-to-ground short circuits by matrix method, *Trans. AIEE*, Vol. 79, Part 3, (1960), p. 323.

6. **H. E. Brown and J. O. Storry,** Improved method of incorporating mutual couplings in single phase short circuit calculations, *IEEE, PICA Conference Proc.*, (1969), pp. 335–342.

7. **D. K. Reitan and K. C. Kruempel,** Modification of the bus impedance matrix for system changes involving mutual couplings, *Proc. IEEE*, (August 1969), p. 1432.

8. **G. W. Stagg and A. H. El-Abiad,** *Computer Methods in Power System Analysis*, McGraw-Hill, 1968.

6

Power Flow
Solutions

Periodically one must add new generating facilities to a power system for load growth. Kirchmayer has shown that the most economical generator to add to a system is one between 10 and 15% of the total installed generating capacity [1]. Addition of this size generating unit causes some transmission facilities to become overloaded during peak load periods. Many power flow solutions are necessary to eliminate these overloads and minimize the cost of installing additional transmission lines and transformers. The plan must permit system operation without overloading any transmission elements, under normal peak load conditions, and during any severe contingency condition that might occur.

A contingency is the loss of a major transmission element or a large generating unit. A double contingency is the loss of two transmission lines, two generators, or a line and a generator. The degree of multiplicity of the contingency to be used in planning a system is a management decision. The engineering problem is the minimization of the investment required to build a system that will withstand the most severe contingencies of the prescribed degree.

Prior to 1930 all power flow calculations were made by hand. Between 1930 and 1956 network calculators* or network analysers† were used to solve these problems.

These devices are miniature models of the network being studied. The behavior of the system is determined by measuring the electrical quantities in the model. In 1956 Ward and Hale described the first really successful computer program for solving power flow problems [2]. The industry converted very rapidly to the computer. The 50 network calculators that

---

* Westinghouse.
† General Electric

had been installed at manufacturers' headquarters, educational institutions, and large utility companies gradually disappeared.

The first method developed for the solution of the equations that describe an electrical network was the Gauss-Seidel algorithm for the solution of linear equations. Since the network equations are quadratic, an iterative procedure is required. Because of the nature of the parameters in a power system network, a solution is usually obtained. The increase in the number of high voltage interconnections in the 1960s caused the number of buses used in representing a system to grow very rapidly. The Gauss-Seidel method encountered greater and greater difficulty in arriving at a solution of the larger networks. Since, as will be seen, the effect of an adjustment in the bus voltage during an iteration is reflected only to the buses that are the immediate neighboring buses, several iterations are required for the adjustments to propagate across the system. Conflicting adjustments may be being made and the number of iterations increased dramatically for large systems. In some cases no solution was obtained for an actually workable system.

A very successful method of calculation of power flows resulted from several years of research by the Bonneville Power Administration [3–6]. This method uses the Newton-Raphson algorithm for solving the simultaneous quadratic equations that describe the power system. The number of iterations required to obtain a solution is practically independent of the size of the system. Many cases that could not be solved by the Gauss-Seidel method, notably systems with negative impedances, are solved successfully by this method. The algorithm is somewhat more susceptible to failure than are other methods, if the starting values of the voltage profile are not judiciously choosen. Early programs [6] used one iteration of Gauss-Seidel before beginning the Newton-Raphson procedure. This process was found to be ill-advised because the Gauss algorithm usually distorts the voltage profile on the first few iterations and causes some buses of the system to be further from a solution than is the original estimate. In addition the restriction of nonnegative impedances imposed by the Gauss method unnecessarily limits the Newton algorithm.

The Jacobian matrix of the Newton method requires considerably more memory than the Y-matrix of the Gauss-Seidel method, but the Newton-Raphson technique is so successful that most of the programs now being written use this algorithm.

Another power flow algorithm that has been developed that has good convergence characteristics is the Z-matrix method [7]. It has the disadvantage of requiring a very large computer memory, since the Z-matrix is full and not sparse as is the case of the Y-matrix and the Jacobian matrix. However, by resorting to diakoptics large systems can be solved using the

Z-matrix method [8–10]. The computer program is more complex, and since there is, as yet, no efficient computer method of "tearing" the system, the method has not been used extensively. An outgrowth of the method is a high-speed method of contingency evaluation, which is discussed in Chapter 7.

All three methods are described here; the Gauss-Seidel because of its historical significance and the ease with which it can be programmed; the Newton-Raphson because of its present popularity in the industry; and the Z-matrix because of its extension to contingency evaluation. The Gauss-Seidel and Z-matrix are discussed only briefly because of their restricted usage. The Newton-Raphson is not described in detail, since an entire volume would be required to do justice to the programs that have been and are being written to implement the algorithm. Rather the reader is directed to the considerable source material [3–6, 11–14, 16].

GAUSS-SEIDEL METHOD

The Gauss-Seidel method of solution developed rapidly because of the ease of writing a program to implement the algorithm. Furthermore, the computer memory requirement of the method is minimal. This factor was an important consideration when load flow programs were being written for systems of 99 buses using computers with 2000 words of memory [15]. Although the method has now largely been replaced, it is discussed briefly because of its historical significance. In addition the algorithm is a good introduction to iterative methods. A basic program can be written by students in a reasonable amount of time as a project assignment.

Each node of a network has four variable quantities associated with it. These quantities are voltage magnitude, voltage angle, real power, and reactive power being supplied to the bus. At each node of the network two of these quantities are prescribed and two are to be determined. There are three types of nodes:

1. A bus with a fixed P (real power) and Q (reactive power) to be supplied to a customer or substation. E (voltage magnitude) and Θ (voltage angle) measured with respect to a fixed reference are to be determined.

2. A generator bus that supplies a fixed P at a given E for which Q and Θ must be determined.

3. A fixed reference (swing bus or slack bus) for which E and Θ are given, and P, Q are to be determined. This bus must supply the difference between the sum of the P's of bus type 1 and the sum of the P's of bus type 2 plus the I^2R losses of the network. It must also supply the sum of the Q's of the other buses plus the I^2X losses of the network.

A solution of the network has been obtained when a voltage profile (voltage magnitudes and angles) is determined that satisfies the imposed constraints. The number of equations that must be satisfied are as follows. Two equations must be satisfied for each fixed P and Q bus in which the E and Θ are the unknowns to be determined. One equation for a fixed E and P bus in which Θ is the unknown. There are no equations required for the fixed E and Θ bus.

Once all the voltages and angles have been determined all the other quantities describing the network can be determined. The solution begins with an estimate of the voltage profile. By an iterative process the original estimate is refined. When the adjustments being made to the voltage and angle during an iteration fall below a prescribed amount on every bus, the voltage profile is accepted and the flow in lines, generated vars, and such are computed and included in the output of the load flow case.

To illustrate the technique used in carrying out the calculation involved in the iterative process, consider a 3-bus portion of a large system as shown in Fig. 6.1.

The bus voltages are assigned starting values that become input to the program. Usually the fixed load buses are assigned $E = 1.0$ and $\Theta = 0°$. The fixed voltage buses, including the swing bus, are assigned the desired value of voltage and an angular position of zero degrees. The network impedances, in per unit on a common voltage and MVA base, are also part of the input to the program. The impedances are converted by the program to admittances and usually expanded to a double entry list. That is, the line data list is expanded so that a line from bus A to bus B, which was included in the system data input, will also appear in the expanded list as a line from B to A. This expanded admittance data list is sorted numerically by bus numbers but never stored in matrix form.

For the portion of the system given in Fig. 6.1, the line list will indicate a line from bus 1 to bus 2 with an admittance y_{1-2} and a line from bus 1 to bus 3 with an admittance y_{1-3}. Also stored in suitable tables are the original estimates of the voltages E_1, E_2, and E_3 and their angular positions Θ_1, Θ_2, and Θ_3 with respect to the swing bus. The desired $P_1 + jQ_1$ entering the system at bus 1 has been prescribed. Using these quantities the iteration begins.

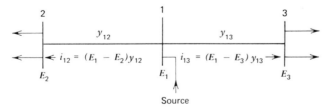

Fig. 6.1.

The current flow from bus 1 to bus 2 is given by

$$I_{1-2} = (E_1 - E_2)y_{12} \qquad (6.1)$$

The power supplied to the bus from an external source is given by

$$E_1 I_1^* = P_1 + jQ_1$$

$$I_1 = \frac{P_1 - jQ_1}{E_1^*} \qquad (6.2)$$

Here the asterisk indicates the complex conjugate. The current I_1 is injected into the bus by a source; a load will therefore have a negative current injection.

The sum of the currents flowing away from bus 1 must be equal to the current injection of the source connected to bus 1.

$$(E_1 - E_2)y_{12} + (E_1 - E_3)y_{13} = \frac{P_1 - jQ_1}{E_1^*} \qquad (6.3)$$

By rearranging and adapting the admittance matrix notation in which $Y_{12} = -y_{12}$, $Y_{13} = -y_{13}$ and $Y_{11} = y_{12} + y_{13} + y_{10}$, one can write this equation:

$$+ E_2 Y_{12} + E_3 Y_{13} - \frac{P_1 - jQ_1}{E_1^*} + E_1 Y_{11} = 0 \qquad (6.4)$$

The self-admittance of bus 1, Y_{11}, is the sum of all the admittances connected to bus 1, including y_{10}, which is the sum of shunt admittances to ground. The terms P_1 and Q_1 are constants, since they represent the power to be supplied to bus 1 by the source. The transfer admittances, Y_{12} and Y_{13}, are constants describing the network. At this point in the iterative solution E_2 and E_3 are considered to be constants. Their values are the results of the last iteration. For the first iteration these values are equal to the original estimate; E_1^* is the conjugate of the voltage on bus 1 as determined on the previous iteration. All values entering equation 6.4 are known or assumed to be fixed except E_1. A revised value E_1' is computed using equation 6.5.

$$E_1' = \frac{- E_2 Y_{12} - E_3 Y_{13} + (P_1 - jQ_1)/E_1^*}{Y_{11}} \qquad (6.5)$$

It is recognized that E_1^* used in equation 6.5 is not the conjugate of this new value of E_1'. The calculation of equation 6.5 is repeated using the conjugate of the new value E_1'.

It has been found that the Gauss-Seidel iterative process can be accelerated by increasing the size of the adjustment made during each iteration.

If E_1' is the new computed value and E_1 was the previous value, the suggested adjustment is given by

$$\Delta E_1 = E_1' - E_1 \tag{6.6}$$

The accelerated value is given by $\Delta E_1' = \alpha \Delta E_1$ where α has been found by experience with power systems to be between 1.2 to 1.6. The new value to be used for the revised value of bus voltage of bus 1 is

$$E_1'' = E_1 + \alpha(E_1' - E_1) \tag{6.7}$$

The iterative process now moves along to bus 2, and all quantities except E_2 in the equation for the sum of the currents flowing away from this bus are considered to be fixed. When a revised voltage has been computed for every bus, the program goes back to the beginning and repeats the process. It is recognized that a new solution for E_1 will be required because E_2 and E_3 have revised values in equation 6.5. When ΔE_n becomes less than a prescribed value (say 0.0001 pu) for every bus, the solution is completed. The line flows and such other information that is required can be computed and printed for analysis purposes.

NEWTON-RAPHSON POWER FLOW

The Newton-Raphson method of solution of a power flow problem was described by Van Ness [3] in 1961. Small test problems demonstrated that the algorithm would solve problems that could not be solved using the Gauss-Seidel method. The technique produced a solution in a very few iterations. It apparently had a speed advantage over the other method, but additional memory was required for storage of the Jacobian matrix over that required for the nodal admittance matrix (double entry line list) of the Gauss algorithm. Because of the ability to solve difficult cases and the apparent speed advantage, the Bonneville Power Administration Group decided to replace the solution link in their power flow program with the Newton-Raphson algorithm [5]. In production work they found that the program now was slower than the Gauss-Seidel method and required a great deal more memory. The Bonneville Group concluded that the difficulty was not with the Newton technique but with the ordering of the equations in the elimination procedure. The remedy was developed by Sato and Tinney and reported in the literature [4]. The memory requirement still exceded that of the Gauss method but the speed and stability of solution justified the conversion to the method.

The number of iterations required by Newton-Raphson to obtain a solution is almost independent of the system size. The restrictions imposed by the Gauss-Seidel method,—that the impedances in the network may not be negative and that there is a limited allowable ratio of individual line admittances to the sum of the admittances connected to a bus—are not as restrictive in Newton's method.

The Newton's method is derived from the Taylor's expansion of a function. A two-dimensional problem is used to review the technique (see Fig. 6.2).

$$f(x_0 + h) = f(x_0) + f'(x_0)h + \frac{f''(x_0)h^2}{2!} + \cdots \tag{6.8}$$

A linear approximation is obtained by dropping all terms beyond the first derivative.

$$f(x_0 + h) \approx f(x_0) + f'(x_0)h \tag{6.9}$$

The suggested increment h that should make $f(x_0 + h)$ approach zero is therefore

$$h \approx -\frac{f(x_0)}{f'(x_0)} \tag{6.10}$$

As can be seen by referring to Fig. 6.2, the linear approximation does not give the required value of x. However, if the initial value x_0 is within a restricted range of the desired root, the process will iterate to a solution. Starting values beyond that range will lead to divergence.

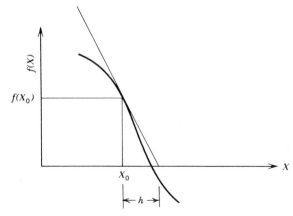

Fig. 6.2. Approximation of a root by linearization.

In the power flow problem, the total current flowing away from bus k over lines connected to neighboring buses is given by equation 6.11.

$$\bar{I}_k = \sum_{m=1}^{N} \bar{Y}_{km} \bar{E}_m \qquad (6.11)$$

where $\bar{E}_m$ is the complex node to datum voltage of bus m with respect to a reference node; $\bar{I}_k$ is the complex current injected into bus k by the generator; $\bar{Y}_{km}$ is the complex admittance matrix element between bus k and bus m. The summation includes all buses of the system that have a direct connection to bus k. The vector quantities can be represented in polar or rectangular form

$$\bar{I}_k = I_k \epsilon^{j\alpha_k} = a_k + jb_k \qquad (6.12)$$

$$\bar{E}_m = E_m \epsilon^{j\delta_m} = e_m + jf_m \qquad (6.13)$$

$$\bar{Y}_{km} = Y_{km} \epsilon^{j\Theta_{km}} = G_{km} + jB_{km} \qquad (6.14)$$

where α and δ are measured with respect to an arbitrary reference voltage and Θ is the angle obtained from the impedance diagram for the particular line between bus k and bus m.

The power delivered to the bus in terms of the injection current $\bar{I}_k$ and the node voltage $\bar{E}_k$ is given by equation 6.15

$$P_k + jQ_k = \bar{E}_k \bar{I}_k^* \qquad (6.15)$$

Note. The asterisk indicates "complex conjugate."

Substitution of the conjugate of $\bar{I}_k$ as given by equation 6.11 into equation 6.15 gives

$$P_k + jQ_k = \bar{E}_k \Sigma \bar{Y}_{km}^* \bar{E}_m^* \qquad (6.16)$$

In a system of N buses there is a set of $N-1$ simultaneous equations of the form of equation 6.16. These equations must be solved simultaneously for the unknown voltages that satisfy the set.

Substitution of the assigned starting values of the bus voltages into equations 6.16 will not give the desired values of $P_k + jQ_k$ as prescribed by the input data. The difference between the desired and the computed quantities $P_k - P_k' = \Delta P_k$ and $Q_k - Q_k' = \Delta Q_k$ must ultimately be reduced below a permissible error by adjustment of the complex voltages $\bar{E}_k$ during the iterative procedure.

The linear approximation of equation 6.9 is extended to the multidimensional case of the load flow problem. Taking total derivatives of equation 6.16 and separating real and imaginary equations 6.17 and 6.18 are obtained.

$$\Delta P_k = \Sigma \frac{\partial P_k}{\partial \delta_m} \Delta \delta_m + \Sigma \frac{\partial P_k}{\partial E_m} \Delta E_m \qquad (6.17)$$

$$\Delta Q_k = \Sigma \frac{\partial Q_k}{\partial \delta_m} \Delta \delta_m + \Sigma \frac{\partial Q_k}{\partial E_m} \Delta E_m \qquad (6.18)$$

In these equations m takes the number k and all bus numbers of those buses directly connected to bus k. The simultaneous set of linear equations, in which k takes all bus numbers except the number of the swing bus, must be solved for the set of values $\Delta \delta$ and ΔE. These adjustments $\Delta \delta$ and ΔE will reduce the errors ΔP and ΔQ. The notation is simplified by writing:

$$\Delta P_k = \Sigma H_{km} \Delta \delta_m + \Sigma N_{km} \frac{\Delta E_m}{E_m} \qquad (6.19)$$

$$\Delta Q_k = \Sigma J_{km} \Delta \delta_m + \Sigma L_{km} \frac{\Delta E_m}{E_m} \qquad (6.20)$$

To explain the method of evaluation of H_{km}, N_{km}, J_{km} and L_{km}, it is convenient to use the polar form, $E_m \epsilon^{j\delta_m}$ for the voltages and the rectangular form for the admittances $G_{km} + jB_{km}$.

$$P_k + jQ_k = \Sigma E_k E_m \epsilon^{j(\delta_k - \delta_m)} (G_{km} - jB_{km}) \qquad (6.21)$$

Two cases arise when taking derivatives $k = m$ and $k \neq m$.

When $k \neq m$ For a particular m, a single term results from the differentiation of (6.21) with respect to δ_m.

$$\frac{\partial P_k}{\partial \delta_m} + j \frac{\partial Q_k}{\partial \delta_m} = -jE_k E_m \epsilon^{j(\delta_k - \delta_m)} (G_{km} - jB_{km}) \qquad (6.22)$$

The single term can be formed using values from the voltage vector of the previous iteration and the single admittance matrix element. Separating real and imaginary parts gives:

$$\frac{\partial P_k}{\partial \delta_m} = \text{Re}\left\{ -jE_k E_m[\cos(\delta_k - \delta_m) + j\sin(\delta_k - \delta_m)](G_{km} - jB_{km}) \right\}$$

$$= E_k E_m[G_{km}\sin(\delta_k - \delta_m) - B_{km}\cos(\delta_k - \delta_m)] \tag{6.23}$$

$$\frac{\partial Q_k}{\partial \delta_m} = \text{Im}\left\{ -jE_k E_m[\cos(\delta_k - \delta_m) + j\sin(\delta_k - \delta_m)](G_{km} - jB_{km}) \right\}$$

$$= -E_k E_m[G_{km}\cos(\delta_k - \delta_m) + B_{km}\sin(\delta_k - \delta_m)] \tag{6.24}$$

Differentiation of (6.21) with respect to E_m gives:

$$\frac{\partial P_k}{\partial E_m} + j\frac{\partial Q_k}{\partial E_m} = E_k \epsilon^{j(\delta_k - \delta_m)}(G_{km} - jB_{km}) \tag{6.25}$$

Multiplying both sides by E_m and separating real and imaginaries gives

$$E_m\frac{\partial P_k}{\partial E_m} = E_k E_m[G_{km}\cos(\delta_k - \delta_m) + B_{km}\sin(\delta_k - \delta_m)] \tag{6.26}$$

$$E_m\frac{\partial Q_k}{\partial E_m} = E_k E_m[G_{km}\sin(\delta_k - \delta_m) - B_{km}\cos(\delta_k - \delta_m)] \tag{6.27}$$

Therefore

$$H_{km} = L_{km} \tag{6.28}$$

and

$$J_{km} = -N_{km} \tag{6.29}$$

FOR THE CASE $m = k$

Differentiation of equation 6.21 with respect to δ_k must be carried out as the derivative of a product.

$$\frac{\partial P_k}{\partial \delta_k} + j\frac{\partial Q_k}{\partial \delta_k} = jE_k \epsilon^{j\delta_k}\Sigma E_m \epsilon^{-j\delta_m}(G_{km} - jB_{km})$$

$$+ E_k \epsilon^{j\delta_k}[-jE_k \epsilon^{-j\delta_k}(G_{kk} - jB_{kk})] \tag{6.30}$$

This equation can be simplified to

$$\frac{\partial P_k}{\partial \delta_k} + j \frac{\partial Q}{\partial \delta k} = j P_k - Q_k - E_k^2 B_{kk} - j E_k^2 G_{kk} \tag{6.31}$$

Separating reals and imaginaries gives

$$H_{kk} = -Q_k - B_{kk} E_k^2 \tag{6.32}$$

$$J_{kk} = P_k - G_k K E_k^2 \tag{6.33}$$

Differentiation of equation 6.21 with respect to E_k must also be carried out as the derivative of a product.

$$\frac{\partial p_k}{\partial E_k} + j \frac{\partial Q_k}{\partial E_k} = \epsilon^{j\delta_k} \Sigma E_m \epsilon^{-j\delta_m} (G_{km} - j B_{km})$$

$$+ E_k \epsilon^{j\delta_k} \epsilon^{-j\delta_k} (G_{kk} - j B_{kk}) \tag{6.34}$$

Multiplication of both sides by E_k gives

$$E_k \left(\frac{\partial P_k}{\partial E_k} + j \frac{\partial Q_k}{\partial E_k} \right) = [E_k \epsilon^{j\delta_k} \Sigma E_m \epsilon^{-j\delta_m} (G_{km} - j B_{km})$$

$$+ E_k^2 (G_{kk} - j B_{kk})] \tag{6.35}$$

Separating reals and imaginaries gives

$$E_k \frac{\partial P_k}{\partial E_k} = P_k + G_{kk} E_k^2 = N_{kk} \tag{6.36}$$

$$E_k \frac{\partial Q_k}{\partial E_k} = Q_k - B_{kk} E_k^2 = L_{kk} \tag{6.37}$$

Note: For the case that k is a generator, Q_k is not specified but is obtained from equation 6.21 by discarding the real parts,

$$Q_k = \mathrm{Im} \Sigma E_k E_m \epsilon^{j(\delta_k - \delta_m)} (G_{km} - j B_{km}) = \Sigma E_k E_m (G_{km} \sin(\delta_k - \delta_m) - B_{km} \cos(\delta_k - \delta_m)).$$

This value of Q_k is used in equation 6.25 if it is in the range, $Q_{k\,min} < Q_k < Q_{k\,max}$. If Q_k is outside the permissible range, bus k is converted to a fixed P, Q bus and the limit that has been exceeded becomes the desired Q_k.

FORMING THE JACOBIAN MATRIX

Using the values for H, N, J, and L for the various buses one forms the Jacobian matrix:

$$\left[\frac{\Delta P}{\Delta Q} \right] = \left[\begin{array}{c|c} H & N \\ \hline J & L \end{array} \right] \left[\frac{\Delta\delta}{\frac{\Delta E}{E}} \right] \tag{6.38}$$

This matrix relates the linearized relationship between small changes in the bus voltage angular positions $\Delta\delta$ and small changes in the bus voltage magnitudes $\Delta E/E$ to changes in bus real and reactive power. The solution of these simultaneous equations give values of $\Delta\delta$ and $\Delta E/E$ that would reduce ΔP and ΔQ to zero if the bus powers are linear relationships of bus voltage magnitude and angle but the equations are quadratic and therefore an iterative process is necessary. New values of power delivered to the buses must be computed using the revised values of δ and $|E|$ in equation 6.16. The new differences between actual power and the desired power define new values of ΔP and ΔQ, and the process begins a new iteration.

In the assembly of the Jacobian matrix of equation 6.38 sparcity must be achieved. Also every effort must be made to obtain a near optimum order for incorporating the buses into the matrix to achieve the speed and economy of memory that makes this method so successful.

The Jacobian matrix is not assembled and then triangularized by the Gaussian elimination, rather the matrix is assembled in the triangularized form and only nonzero elements are stored in the matrix. This is a very important consideration in reducing the memory requirement for a given network.

The details of how this is done are explained in terms of the bus admittance matrix, since it contains only single entries between the buses that are connected, whereas the Jacobian matrix has one, two, or four entries depending on the types of buses being connected. If bus k is a fixed E and fixed P bus, equation 6.20 does not apply, since Q has not been specified. If in addition, bus m is a fixed voltage bus, then $\Delta E_m = 0$ because E_m is not allowed to vary and therefore N_{km} will not be required. In this case, k and m are both fixed voltage buses, and H_{km} is the only entry in the Jacobian matrix. If k is a fixed E bus and m is a fixed P and Q bus, there will be two entries, H_{km} and N_{km}. Finally if k and m are both fixed P, Q buses, all four quantities H_{km}, N_{km}, J_{km}, and L_{km} must be entered in the matrix.

To eliminate the complication of different number of entries, which would only confuse the explanation of the process, that is, forming the Jacobian matrix in triangular form, the simple admittance matrix is used.

The first rows incorporated into the matrix correspond to buses with the least number of connections. This keeps the amount of "fill-in" at a near minimum during the Gaussian elimination. Thus advantage is taken of sparcity of the matrix both in the amount of computation that must be made and also in the memory requirement for storage of the matrix. As buses are added to the matrix, the number of connections to remaining unprocessed buses is adjusted for the lines included in the matrix in adding the row for the bus being processed. The Ward and Hale network is used to illustrate formation of the matrix [2] (see Fig. 6.3).

The number of connections to each bus are given in Table 6.1.

Table 6.1

Bus Number	Number of Lines Connected	Connecting Buses
3	2	2,4
1	2	4,6
5	2	2,6
2	2	3,5
6	3	1,4,5
4	3	1,3,6

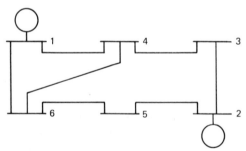

Fig. 6.3.

Forming the matrix should begin with one of the buses that has only two lines connected. Bus 3 is satisfactory. The first row of the admittance matrix would then be given in Fig. 6.4.

$$3 \qquad 2 \qquad 4$$
$$3 \quad [\Sigma Y_3 \quad -Y_{32} \quad -Y_{34}] \qquad \textbf{Fig. 6.4.} \text{ Matrix entries after including bus 3.}$$

The lines and connections of the remaining unprocessed buses are given in Table 6.2

Table 6.2

Bus Number	Number of Connecting Lines	Unprocessed Connections
1	2	4,6
5	2	2,6
2	1	5
6	3	1,4,5
4	2	1,6

Bus 2 should now be processed, since it would introduce only one column in the matrix. The two rows of the matrix would now be given in Fig. 6.5.

$$
\begin{array}{cccc}
3 & 2 & 4 & 5 \\
\end{array}
$$
$$
\begin{array}{c}
3 \\
2
\end{array}
\left[
\begin{array}{cccc}
\Sigma Y_3 & -Y_{32} & -Y_{34} & 0 \\
-Y_{23} & \Sigma Y_2 & 0 & -Y_{25}
\end{array}
\right]
$$

Fig. 6.5. Matrix entries after including buses 3 and 2.

Only the nonzero elements will be stored in memory. The beginning and end of rows and the buses being connected are recorded in lists. The zero elements in Fig. 6.5 and Fig. 6.6 are shown but would not be stored in the matrix. The table of lines and connections is corrected for the addition of bus 2 (see Table 6.3).

Table 6.3

Bus Number	Number of Lines Connecting	Unprocessed Connections
1	2	4,6
5	1	6
6	3	1,4,5
4	2	1,6

Bus 5 now has the least number of connections and should be processed next. If the bus admittance matrix is completed using this process, it would be as shown in Fig. 6.6.

$$
\begin{array}{cc}
& \begin{array}{cccccc} 3 & \quad 2 & \quad 5 & \quad 6 & \quad 1 & \quad 4 \end{array} \\
\begin{array}{c} 3 \\ 2 \\ 5 \\ 6 \\ 1 \\ 4 \end{array} &
\left[
\begin{array}{cccccc}
\Sigma Y_3 & -Y_{32} & 0 & 0 & 0 & -Y_{34} \\
-Y_{23} & \Sigma Y_2 & -Y_{25} & 0 & 0 & 0 \\
0 & -Y_{52} & \Sigma Y_5 & -Y_{56} & 0 & 0 \\
0 & 0 & -Y_{65} & \Sigma Y_6 & -Y_{61} & -Y_{64} \\
0 & 0 & 0 & -Y_{16} & \Sigma Y_1 & -Y_{14} \\
-Y_{43} & 0 & 0 & -Y_{46} & -Y_{41} & \Sigma Y_4
\end{array}
\right]
\end{array}
$$

Fig. 6.6. The untriangularized matrix without exploiting sparcity.

In the formation of the matrix the Gaussian elimination process is carried out as the rows are being added. Row 1 of the matrix of Fig. 6.4 is divided by ΣY_3 to give the row shown in Fig. 6.7. Modified elements are shown by prime marks.

$$
\begin{bmatrix} 1 & -Y_{32}' & 0 & 0 & 0 & -Y_{34}' \end{bmatrix}
$$

Fig. 6.7. The beginning of the Gaussian elimination.

This row is multiplied by $-Y_{23}$ and subtracted from row 2 of Fig. 6.6 to give the first two rows shown in Fig. 6.8.

$$
\begin{bmatrix}
1 & -Y_{32}' & 0 & 0 & 0 & -Y_{34}' \\
0 & \Sigma Y_2' & -Y_{25}' & 0 & 0 & -Y_{24}'
\end{bmatrix}
$$

Fig. 6.8. The element below the diagonal has been eliminated.

Note the fill-in that has occurred in Y_{24}' in Fig. 6.8.

Before being stored in the matrix, row 2 is divided by $\Sigma Y_2'$ and the first two rows are given in Fig. 6.9.

$$
\begin{bmatrix}
1 & -Y_{32}' & 0 & 0 & 0 & -Y_{34}' \\
 & 1 & -Y_{25}'' & 0 & 0 & -Y_{24}''
\end{bmatrix}
$$

Fig. 6.9. The elimination continues without exploiting sparcity.

Sufficient terms have been determined to illustrate the process. After the Gaussian elimination has been completed, all diagonal elements are unity and need not be stored. Eliminate all unity and zero elements from the first two rows. The remaining elements are stored sequentially in memory as shown in Fig. 6.10.

$$[\, Y'_{32} \quad Y'_{34} \quad Y''_{25} \quad Y''_{34} \,]$$

Fig. 6.10. Exploiting sparcity in storing rows for buses 3 and 2.

In Fig. 6.10 the first 12 elements of the matrix of Fig. 6.6 have been compressed into four elements. The actual location of the elements and the buses being connected are kept track of by lists and pointers.

Expansion of this technique to the Jacobian matrix is illustrated. The Jacobian matrix for the Ward and Hale network is shown in Fig. 6.11. Bus 1 is the swing bus and does not have an entry, since both δ_1 and E_1 are fixed. Bus 2 is a fixed E and P bus. All other buses are fixed P and Q buses.

$$
\begin{bmatrix}
\Delta P_3 \\
\Delta Q_3 \\
\Delta P_2 \\
\Delta P_5 \\
\Delta Q_5 \\
\Delta P_6 \\
\Delta Q_6 \\
\Delta P_4 \\
\Delta Q_4
\end{bmatrix}
=
\begin{bmatrix}
H_{33}N_{33} & H_{32} & 0 & 0 & 0 & 0 & H_{34}N_{34} \\
J_{33}L_{33} & J_{32} & 0 & 0 & 0 & 0 & J_{34}L_{34} \\
H_{23}N_{23} & H_{22} & H_{25}N_{25} & 0 & 0 & 0 & 0 \\
0 & 0 & H_{25} & H_{55}N_{55} & H_{56}N_{56} & 0 & 0 \\
0 & 0 & J_{25} & J_{55}L_{55} & J_{56}L_{56} & 0 & 0 \\
0 & 0 & 0 & H_{65}N_{65} & H_{66}N_{66} & H_{64}N_{64} \\
0 & 0 & 0 & J_{65}L_{65} & J_{66}L_{66} & J_{64}L_{64} \\
H_{43}N_{43} & 0 & 0 & 0 & H_{46}N_{46} & H_{44}N_{44} \\
J_{43}L_{43} & 0 & 0 & 0 & J_{46}L_{46} & J_{44}L_{44}
\end{bmatrix}
\begin{bmatrix}
\Delta\delta_3 \\[2pt]
\dfrac{\Delta E_3}{E_3} \\[6pt]
\Delta\delta_2 \\[2pt]
\Delta\delta_5 \\[2pt]
\dfrac{\Delta E_5}{E_5} \\[6pt]
\Delta\delta_6 \\[2pt]
\dfrac{\Delta E_6}{E_6} \\[6pt]
\Delta\delta_4 \\[2pt]
\dfrac{\Delta E_4}{E_4}
\end{bmatrix}
$$

Fig. 6.11. The complete Jacobian matrix of the system of fig. 6.3.

Note that although no equation is given for bus 1, its effect on the system is taken into account in computing $P_6 + jQ_6$ and $P_4 + jQ_4$ by equation 6.16. It must also be remembered that the matrix is not stored as shown in Fig.

6.11 but is triangularized and stored row by row. The items for this matrix would be stored sequentially as:

$$N'_{33}H'_{32}H'_{34}N'_{34}J'_{32}J'_{34}L'_{34}H'_{25}N'_{25}H'_{24}N''_{55}H''_{56}N''_{56}\ldots$$

Elements J_{33}, H_{23}, and N_{23} will have been reduced to zero and not stored. Elements H_{33}, L_{33}, and H_{22} will have been converted to unity and not stored. The last row of the matrix will have been reduced to

$$\Delta Q'_4 = L'_{44}\left(\frac{\Delta E_4}{E_4}\right) \tag{6.39}$$

From this equation, ΔE_4 can be determined. Its value is used in the back substitution into the equation from the second last row of the triangularized Jacobian matrix.

$$\Delta P'_4 = H'_{44}\Delta\delta_4 + N'_{44}\frac{\Delta E_4}{E_4} \tag{6.40}$$

This equation is solved for $\Delta\delta_4$, and the back substitution continued until the vector of corrections has been completed. The bus voltage vector is corrected, and the next iteration begins.

The Jacobian matrix must be reformed for each iteration, because buses can change types as a result of limit checking during an iteration. For example, a fixed voltage bus may not be able to hold its voltage to the desired value because of the maximum (minimum) var limitation prescribed by the input data. If a generator exceeds its var limit, the generator is changed to a fixed P and Q bus and the Q is set back to the var limit that was violated. In the next iteration due account must be taken of this change of bus type in formation of the matrix. In later iterations the bus may revert to its original type if the var limit and desired voltage again come within range.

REFINEMENTS IN THE NEWTON'S METHOD

After a solution had been obtained in problems in which an interchange schedule between members of a pool operation had been specified, it originally was necessary to sum up the flow in the tie lines between neighbors. The sum of the flow on the tie lines was compared with the intended interchange. If the actual interchange did not satisfy the requirement within a prescribed tolerance, generation of alternate swing generators in the individual areas were adjusted and the solution repeated with the *fixed P* of certain generators changed to new values. Britton [12] pointed out that the alternate swing generators were really not fixed P

buses, but that the sum of the flows on the tie lines was the constraint to be met. He suggested that the equation for the alternate swing generator be replaced by an equation for the sum of the tie line flows. This was a definite improvement, since the first solution of the network provided the required interchange. A considerable savings in computer running time resulted.

The regulation of a bus voltage by a tap changer and the regulation of the Mw flow in a line by adjustment of a phase shifter experienced the same difficulty. After a solution had been obtained, the actual quantity was compared with the desired value. If it was not within the tolerance, the tap changer (or phase shifter) was adjusted and the solution repeated as often as was necessary.

Simultaneously Britton [17] and the team of Peterson and Meyer [13] solved this problem by making the complex tap a variable and by introducing an appropriate equation into the Jacobian matrix equation. Care must be exercised in entering this equation to ensure that the diagonal term is nonzero at the time when division by it is required in the Gaussian elimination. The near optimal ordering of the equations may be slightly disturbed, since a particular equation must not be entered until the diagonal term is ensured of being nonzero. The elimination of repetitive solving of a large network to arrive at the desired constraints justifies the method.

A method of assuring a starting voltage profile that will guarantee convergence (if a solution exists) was described by Stott [16]. In this method two DC solutions are obtained by use of reduced Jacobian matrices. In the one solution only voltage magnitudes are permitted to vary and in the other only the angular position of the bus voltages.

THE BUS IMPEDANCE MATRIX POWER FLOW

The success of the Z-bus matrix in the calculation of short circuits as reported in Ref. 3 of Chapter 3 led to research in the use of the impedance matrix in the calculation of power flow problems. In short-circuit work it was very natural to select the common bus behind the generator reactances as the reference bus and to compute the network solution as a DC problem with a single source. In power flow work this is no longer possible because of the different angular position of the various generators. Either the swing bus or the common ground bus is selected as the reference bus [7]. In this discussion the ground bus is selected as the reference bus.

The Z-matrix is assembled one line at a time as was explained in Chapter 3 but with one difference—the matrix is assembled in complex form, $Z = R + jX$. All mathematical operations must be carried out using complex arithmetic, and the real and imaginary parts of the matrix must be stored in separate matrices.

Included with the line data are approximate load impedances to ground determined on the basis of unity bus voltage at an angle of zero using the equation

$$Z = \frac{E^2}{P - jQ} \tag{6.41}$$

A bus injection current vector is formed. At the beginning of the initial iteration, the load current injections are all equal to zero, since the load impedances consume the correct real and reactive power as determined by equation 6.41 for the assumed bus voltage of $E = 1.0 \angle 0°$. The generators inject current into their buses as determined by

$$I = \frac{P - jQ}{E^*} \tag{6.42}$$

The matrix equation for determining bus voltages can then be written

$$ZI = E \tag{6.43}$$

Assume that the first equation of the matrix equation 6.43 is a regulated voltage bus. In expanded form this equation is

$$Z_{11}I_1 + Z_{12}I_2 + Z_{13}I_3 + \cdots + Z_{1S}I_S + \cdots + Z_{1n}I_n = E_1 \tag{6.43'}$$

If the absolute value of the voltage that is obtained for E_1 is not the desired magnitude, the current injection into bus 1 is given an increment ΔI_1 to correct $|E_1|$ to its desired value. The increment ΔI_1 must also satisfy the additional constraint that

$$\text{Re}\{E_1(I_1 + \Delta I_1)^*\} = P_1$$

This increment of current will disturb the voltage E_S of the swing bus which is not allowed to change its magnitude or angle. Therefore simultaneously with the addition of the increment ΔI_1 added to I_1 a counter adjustment must be made to I_S by the addition of ΔI_S to maintain E_S at its desired value.

The original value of swing bus voltage is given by

$$Z_{S1}I_1 + Z_{S2}I_2 + \cdots + Z_{SS}I_S + \cdots Z_{Sn}I_n = E_S \tag{6.44}$$

Simultaneous adjustment ΔI_1 and ΔI_S are made such that E_S remains unchanged.

$$Z_{S1}(I_1 + \Delta I_1) + Z_{S2}I_2 + \cdots Z_{SS}(I_S + \Delta I_S) + \cdots Z_{Sn}I_n = E_S \tag{6.45}$$

Subtraction of equation 6.44 from 6.45 gives

$$Z_{S1}\Delta I_1 + Z_{SS}\Delta I_S = 0 \tag{6.46}$$

$$\Delta I_S = -\frac{Z_{S1}\Delta I_1}{Z_{SS}} \tag{6.47}$$

The driving point impedance of a bus is generally greater than its transfer impedances,

$$Z_{SS} > Z_{S1}$$

consequently

$$\Delta I_S < \Delta I_1$$

The counter adjustment produced by ΔI_S in equation 6.44 will be further reduced, since $Z_{11} > Z_{1S}$ and therefore

$$Z_{11}\Delta I_1 \gg Z_{S1}\Delta I_S$$

In general the change in current injected into a bus will produce much smaller changes in the voltage of all other buses than in its own and the process will converge to a solution in a very few iterations. The more loosely tied systems will converge to a solution with less iterations than do tightly knit systems.

For buses with fixed $P + jQ$ a "fringing current" must be added to the corresponding current of the current vector to maintain the prescribed values of P and Q if E is different than the original value of $E = 1.0 \angle 0°$.

Suppose bus k is a fixed P and Q bus.

$$Z_{k1}I_1 + Z_{k2}I_2 + \cdots Z_{kk}I_k + \cdots Z_{kn}I_n = E_k \tag{6.48}$$

The power consumed by the impedance between bus k and ground when E_k is applied across it is

$$P_k - Q_k = \frac{E_k^2}{Z_{\text{line }(k-\text{ground})}} \tag{6.49}$$

If E_k is different from $E_K = 1.0 \angle 0°$ used in determining $Z_{\text{line }(k-\text{ground})}$, the power $P_k + jQ_k$ will not be the desired amount. The difference between the desired and actual complex power is given by

$$(\overline{P}_k + j\overline{Q}_k) - (P_k + jQ_k) = \Delta P_k + j\Delta Q_k \tag{6.50}$$

A fringing current ΔI_k must be added to I_k in the current vector as given by

$$E_k \Delta I_k^* = \Delta P_k + j\Delta Q_k \qquad (6.51)$$

Again a counter adjustment must be made in the swing bus injection current to maintain its prescribed value. The iteration process continues until $\Sigma|\Delta I_S|$ in a given iteration falls below the allowable total mismatch.

The method has excellent convergence characteristics, is not sensitive to initial values of the voltage profile, and can process negative impedances (series compensation of lines) and zero impedance circuit breakers. It is used to some extent in Europe and to a limited extent in the United States. For systems whose Z-matrix will not fit in core storage the method of diakoptics is successful [10].

References

1. **L. K. Kirchmayer, A. G. Mellor, J. F. O'Mara, and J. R. Stevenson,** An investigation of the economic size of steam electric generating units, *Trans. AIEE*, Vol. 74, Part 3 (August 1955), p. 600.

2. **J. B. Ward and H. W. Hale,** Digital computer solution of power flow problems. *Trans. AIEE*, Vol. 75, Part 3, (June 1956), pp. 398–404.

3. **J. E. Van Ness and J. H. Griffin,** Elimination methods for load flow studies, *Trans. AIEE*, Vol. 80, Part 3, (1961), p. 299.

4. **N. Sato and W. F. Tinney,** Technique for exploiting the sparsity of the network admittance matrix, *Trans. IEEE, PA&S*, Vol. 82, (December 1963), p. 944.

5. **W. F. Tinney and C. E. Hart,** Power flow solutions by Newton's method, *Trans. IEEE, PA&S*, Vol. 86, (November 1967), p. 1449.

6. **W. F. Tinney and J. W. Walker,** Direct solution of sparse network equations by optimally ordered triangular factorization, *Proc. IEEE*, Vol. 55, (November 1967), pp. 1801–1809.

7. **H. E. Brown, G. K. Carter, H. H. Happ, and C. E. Person,** Power flow solution by impedance matrix method, *Trans. AIEE*, Vol. 82, Part 3, (1963), p. 1.

8. **G. Kron,** Diakoptics-The piecewise solution of large scale systems, General Electric Laboratory Report 57GL330, General Electric Co., 1957.

9. **A. Brameller,** The application of diakoptics to network analysis, *Proc. Power Systems Computer Conf.*, Report 4.6, Stockholm 1966.

10. **R. G. Andretich, H. E. Brown, H. H. Happ, and C. E. Person,** The piecewise solution of the impedance matrix load flow, *IEEE Trans. PA&S*, Vol. 87, (October 1968), pp. 1877–1882.

11. **J. Carpentier,** Application of Newton's method to load flow problems, *Proc. Power Systems Computer Conf.*, London, 1963.

12. **J. P. Britton**, Improved area interchange control for Newton's method load flows, *IEEE Trans. PA&S*, (1969), pp. 1577–1579.

13. **N. M. Peterson and W. S. Meyer**, Automatic adjustment of transformer and phase shifter taps in the Newton power flow, *IEEE Trans. PA&S*, Vol. 90, (January-February 1971), pp. 103–108.

14. **S. T. Despotovic, B. S. Babic', and V. P. Mastilovic'**, A rapid and reliable method for solving load flow problems, *Trans. IEEE PA&S*, Vol. 90, (January 1971), p. 123.

15. **R. J. Brown and W. F. Tinney**, Digital solution for large power networks, *Trans. AIEE*, Vol. 76, Part 3, (1957), pp. 347–355.

16. **B. Stott**, Effective starting process for Newton-Raphson load flows, *Trans. IEE, Power*, Vol. 118, London.

17. **J. P. Britton**, Improved load flow performance through a more generalized equation form, *IEEE, Trans. PA&S*, Vol. 90, (January-February 1971), pp. 109–115.

18. **A. F. Glimn and G. W. Stagg**, Automatic calculation of load flows, *Trans AIEE*, Vol. 76, Part 3, (1957), p. 817.

7

High-Speed Reduced Accuracy
Power Flow Calculations
for Contingency Evaluation
and Maximum Interchange
Capability Determinations

We saw in Chapter 6 that many power flow calculations must be made to make sure that a proposed expansion of a power system will not overload transmission lines during multiple contingency conditions. During a contingency condition in a system, especially if there is a generation loss, large amounts of power must be transferred between neighboring companies. The maximum amount of power that can be transferred between members of the interconnection should therefore be known. The testing of expansion plans of large interconnections and the determination of emergency power transfer capability require much computer time. Several factors are responsible for the extensive computer usage in planning and operation studies:

1. A steady increase in the number of buses required to represent properly an interconnected system because of the growing number of EHV interconnections.

2. An increase in computer running time per case as the number of buses in the representation becomes larger.

3. A proliferation of cases that must be studied as the systems become more complex.

There is a great incentive to find methods of evaluating contingencies and to determine the maximum allowable interchange in interconnection planning studies. Faster power flow methods also have an application in real-time on-line system monitoring by security computers located in the operators office.

The desire to increase the speed of computation of power flow cases (decrease the computer running time) attracted the attention of several groups of investigators. The problem of scheduling interchange during contingency conditions was first investigated by MacArthur [1]. El-Abiad and Stagg described a method, using the nodal impedance matrix, for determining the lines that would be overloaded as each line of the system would be opened one at a time. Only single contingencies were studied and the size of the system was limited by the Z-matrix that could be stored in the computer memory [2]. Limmer discussed a method of evaluating multiple contingencies by computing distribution factors for the multiple line outage cases from single outage distribution factors [3]. In this method the distribution factors are computed from elements of the Z-matrix, and the system size is again limited by the available memory of the computer. Brown developed a Z-matrix method that employs a small computer for single contingency evaluations of large systems. The method was originally intended for on-line use by system operators [4]. Later the method was expanded to include multiple contingencies and interchange evaluation for use in system planning and operating studies for very large interconnections. In this method the system size is not limited by the available memory [5,6]. The base case power flow that had been obtained by the Newton-Raphson (or any other method) is retrieved from the computer library, and the analysis of the contingencies is achieved at high speed using the Z-matrix. Peterson, Despotovic, and Stott each developed a faster version of the Newton-Raphson power flow algorithm [7–9]. The Z-matrix method and the decoupled Newton-Raphson methods are described in this chapter.

AXIS DISCARDING IN Z-MATRIX METHODS

The axis discarding technique that extended the capability of the Z-matrix short-circuit program was described in Chapter 4 [10]. In this application the method was completely rigorous because, for each short-circuit condition, the network is subjected to a single current injection. When bus k is in short-circuit condition, only column k of the Z-matrix and the single nonzero current I_k are required in the calculation of line flows (see equation 7.1).

$$
\begin{bmatrix}
Z_{11} & Z_{12} & \cdots & Z_{1k} & \cdots & Z_{1n} & \cdots \\
Z_{21} & Z_{22} & \cdots & Z_{2k} & \cdots & Z_{2n} & \cdots \\
 & & \cdots & & & & \\
 & & \cdots & & & & \\
 & & \cdots & & & & \\
Z_{k1} & Z_{k2} & \cdots & Z_{kk} & \cdots & Z_{kn} & \cdots \\
 & & \cdots & & & & \\
 & & \cdots & & & & \\
 & & \cdots & & & & \\
Z_{n1} & Z_{n2} & \cdots & Z_{nk} & \cdots & Z_{nn} & \cdots \\
 & & \cdots & & & & \\
 & & \cdots & & & & \\
 & & \cdots & & & &
\end{bmatrix}
\begin{bmatrix}
0 \\ 0 \\ 0 \\ 0 \\ 0 \\ I_k \\ 0 \\ 0 \\ 0 \\ 0 \\ 0 \\ 0 \\ 0
\end{bmatrix}
=
\begin{bmatrix}
E_1 \\ E_2 \\ \vdots \\ E_k \\ \vdots \\ E_n \\ \vdots
\end{bmatrix}
\tag{7.1}
$$

Elimination of an axis n by deleting row n and column n from the Z-matrix requires that I_n also be deleted from the current vector to maintain conformability between the Z-matrix and the current vector. Since $I_n = 0$, when bus k is in short-circuit condition, the remaining voltages in the voltage vector are undisturbed by the omission of the axis n from the Z-matrix and the deletion of the zero current I_n from the current vector. Therefore, in fault study work, small portions of the Z-matrix can be retained, and for each single injection current the bus voltages are correct for the portion of the system that is being studied.

In any system the planning engineer and the system operators know, by experience, the trouble spots in their system. In a large interconnected network all of the weaknesses of the entire system are recognized by at least one of the interconnection planning engineers. The critical lines that when opened tend to cause other lines to become overloaded are known. The limiting lines that have a tendency to become overloaded if the system is being operated abnormally are also known. Thus only a small part of the system is of interest or actually required in a contingency evaluation study. In the Z-matrix formulation of the problem it is only necessary to retain the axes corresponding to the critical and limiting lines. Because of the severe memory requirement of the Z-matrix, all other axes that are not required must be discarded as soon as possible during the matrix formation [11]. Thus a very large system can be studied with an amount of memory that depends on only the number of critical and limiting lines being investigated.

Care must be taken that the portion of the Z-matrix that is saved includes all of the axes that will be required for the several contingencies that are to be investigated. This eliminates the necessity of rebuilding the Z-matrix unnecessarily.

It is indicated in Chapter 4 that the Z-matrix contingency evaluation program [4,5] was an outgrowth of the Z-matrix load flow algorithm. It must be emphasized that the contingency program is not an extention of the load flow program but rather is a modification of it. In actual practice [6], the Z-matrix contingency analysis is based on a power flow base case that has been computed using the Newton-Raphson power flow program.

The discarding technique that was so successful in short-circuit work introduces complications when applied to the Z-matrix load flow. Assume that the complete Z-matrix of a system could be retained in memory and that the current vector required for convergence of the case had been determined (see equation 7.2).

$$
\begin{bmatrix}
Z_{11} & Z_{12} & \cdots & Z_{1k} & \cdots & Z_{1n} \\
Z_{21} & Z_{22} & \cdots & Z_{2k} & \cdots & Z_{2n} \\
& & \cdots & & & \\
& & \cdots & & & \\
& & \cdots & & & \\
Z_{k1} & Z_{k2} & \cdots & Z_{kk} & \cdots & Z_{kn} \\
& & \cdots & & & \\
& & \cdots & & & \\
& & \cdots & & & \\
Z_{n1} & Z_{n2} & \cdots & Z_{nk} & \cdots & Z_{nn} \\
& & \cdots & & & \\
& & \cdots & & & \\
& & \cdots & & &
\end{bmatrix}
\begin{bmatrix}
I_1 \\
I_2 \\
\vdots \\
I_k \\
\vdots \\
I_n \\
\vdots
\end{bmatrix}
=
\begin{bmatrix}
E_1 \\
E_2 \\
\vdots \\
E_k \\
\vdots \\
E_n \\
\vdots
\end{bmatrix}
$$

$$(7.2)$$

In equation 7.2 if row k and column k are removed from the Z-matrix, I_k must be removed from the current vector to satisfy the requirement for multiplication that the number of rows in the vector I be equal to the number of columns in the Z-matrix. The E_k is removed from the voltage vector, since it is no longer defined because row k has been removed from the Z-matrix.

The voltages in the vector of equation 7.2 no longer represent the converged load flow conditions if any current I_k has been eliminated. For example, the voltage obtained from equation 7.2 for bus r after axis k has been removed is given by equation 7.3.

$$E_r = \sum_{\substack{i=1 \\ i \neq k}}^{n} Z_{ri} I_i \qquad (7.3)$$

The missing term $Z_{rk} I_k$ in the summation causes E_r to be in error. In the formation of the Z-matrix for contigency analysis a great many axes must be discarded and the corresponding currents deleted from the current vector. The voltages of the buses that remain will be greatly different than the converged power flow voltages.

It follows then that even if a converged Z-matrix power flow was available for the entire system, axis discarding, which is necessary to reduce the problem to a manageable size, would cause the voltage vector to be useless for contingency analysis. The method of obtaining a current vector that yields useful voltages for contingency analysis is discussed in the next topic (p. 129).

The Z-matrix that is assembled for use in power flow calculations included ground tie impedances to approximate the customer load on the basis of 1.0-pu bus voltages. This is only one formulation of the problem. The inadequacy of these ground ties in the power flow program is compensated for by fringing currents injected into the buses [12]. As discussed earlier the voltage vector that results after deleting nonzero currents from the current vector does not give the proper bus voltages for the system. The ground ties therefore are not appropriate for the voltages that result after deleting currents when the Z-matrix is reduced. The transmission system including these ground ties, which no longer match the bus voltages, therefore does not properly represent the network being studied. The equivalent load impedances computed on the basis of 1.0-pu bus voltage should therefore not be included in the transmission network, since axis discarding is being used [5].

Fixed impedances to ground, such as line charging, transformer equivalent ground legs [13], shunt capacitors, and reactors, could properly be included in the network, since they are linear elements. However, this refinement is not warranted within the accuracy requirement in these studies. All ground ties that are used in the Z-matrix load flow have now been shown to be unsuitable or an unnecessary refinement and are therefore stripped from the transmission system. Since ground is to be used as reference, an arbitrary ground tie impedance is connected to a bus that has been selected as the starting point for the formation of the Z-matrix. The axis discarding technique operates most efficiently when the matrix formation begins at a point very remote from the area of interest. The order of the matrix remains small for a longer period in the formation, since buses are added to the system and discarded from the system at about the same rate until buses in the area of study begin to be attached to the system. These axes in the area of study must be saved, and the order of the matrix grows rapidly. The engineer in charge of the study selects the bus to be tied to the reference by an arbitrary ground tie.

THE Z-MATRIX CONTINGENCY EVALUATION METHOD

The line data and the base case converged voltages, as computed by the Newton-Raphson program (or any other power flow algorithm), are available in the computer library. A list of lines that will be opened (or closed) in some combination in the contingencies to be studied, a list of limiting lines that tend to become overloaded, a list of generators that will take part in the interchange capability determination, and the bus to be used as a starting point in the formation of the matrix are prepared and submitted as data to the program. The matrix for the system is computed by means of the building algorithm and stored in auxiliary storage. The matrix that has been saved is shown in Fig. 7.1.

Assume that a single contingency is to be investigated in which line *a-b* is to be opened. Before the line is opened, it must meet the requirement that it is carrying the normal current that the line carried in the base load flow. This requirement will be met if the difference in voltage between the two buses is equal to the voltage difference obtained from the library record of the converged voltages (see Fig. 7.2).

This constraint can be met by injecting a single current of the proper amount into any bus of the system. It is a matter of convenience to inject current into one of the two buses at the end of line *a-b*, the line that is to be removed. A small matrix is extracted from the matrix of Fig. 7.1 and

Fig. 7.1. Z-matrix retained. All other axes of the system have been discarded.

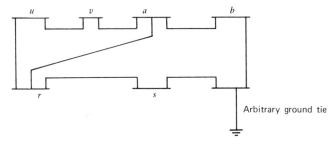

Fig. 7.2. Transmission system with an arbitrary single tie to the reference contingency produced by building a line a to b.

multiplied by the current vector which contains the single nonzero element shown in equation 7.4.

$$a \begin{matrix} a & b \\ \left[\begin{matrix} & Z & \\ & & \end{matrix} \right] \end{matrix} \begin{bmatrix} I_a \\ 0 \end{bmatrix} = \begin{bmatrix} E_a \\ E_b \end{bmatrix} \tag{7.4}$$

This equation can be simplified by subtracting row 2 from row 1 in the Z-matrix and the voltage vector as given in equation 7.5.

$$a\text{-}b \begin{matrix} a & b \\ \left[\begin{matrix} & Z & \\ & & \end{matrix} \right] \end{matrix} \begin{bmatrix} I_a \\ 0 \end{bmatrix} = \begin{bmatrix} E_a - E_b \end{bmatrix} \tag{7.5}$$

The $E_a - E_b$ is the voltage difference across the line $a\text{-}b$ as determined from the base case power flow library record. A further simplification can be made by deleting the second column of the Z-matrix and the corresponding zero current from the current vector.

$$a\text{-}b \, [Z] \, [I_a] = [E_a - E_b] \tag{7.6}$$

or

$$I_a = \frac{E_a - E_b}{Z_{aa} - Z_{ba}} \tag{7.7}$$

To determine the increment of current picked up by lines *r-s* and *u-v*, when line *a-b* is opened, extract Z-matrix elements from the matrix of Fig. 7.1 to form the matrix of Fig. 7.3.

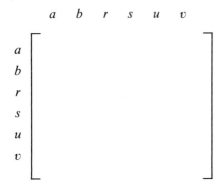

    a   *b*   *r*   *s*   *u*   *v*

Fig. 7.3.   Z-matrix with line *a-b* in the system.

By using the building algorithm add a line between buses *a* and *b* whose impedance is the negative of the original line date. This in effect removes the original line and the matrix of Fig. 7.4 is obtained.

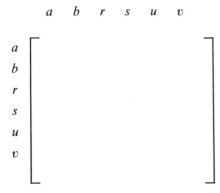

    a   *b*   *r*   *s*   *u*   *v*

Fig. 7.4.   Z'-matrix with the line *a-b* removed.

By extracting the appropriate matrix elements from the matrix of Fig. 7.3, determine the voltage difference across the lines *r-s* and *u-v* caused by the injection of current I_a into bus a (see equation 7.8).

$$
\begin{matrix} & a & b \end{matrix} \\
\begin{matrix} r \\ s \\ u \\ v \end{matrix}
\begin{bmatrix} & & \\ & Z & \\ & & \\ & & \end{bmatrix}
\begin{bmatrix} I_a \\ \\ 0 \\ \\ \end{bmatrix}
=
\begin{bmatrix} E_r \\ E_s \\ E_u \\ E_v \end{bmatrix}
\tag{7.8}
$$

Subtract rows in the Z-matrix and voltages in the vector as was done in obtaining equation 7.5 from equation 7.4.

$$
\begin{matrix} & a & b \end{matrix} \\
\begin{matrix} r\text{-}s \\ \\ u\text{-}v \end{matrix}
\begin{bmatrix} & & \\ & Z & \\ & & \end{bmatrix}
\begin{bmatrix} I_a \\ \\ 0 \end{bmatrix}
=
\begin{bmatrix} E_r - E_s \\ \\ E_u - E_v \end{bmatrix}
\tag{7.9}
$$

Omit column b and the zero current in the current vector to obtain equation 7.10.

$$
\begin{matrix} & a \end{matrix} \\
\begin{matrix} r\text{-}s \\ u\text{-}v \end{matrix}
\begin{bmatrix} \\ Z \\ \end{bmatrix}
\begin{bmatrix} \\ I_a \\ \end{bmatrix}
=
\begin{bmatrix} E_r - E_s \\ E_u - E_v \end{bmatrix}
\tag{7.10}
$$

The current injection can only meet the one requirement that $E_a - E_b$ is correct. The two voltage differences given in equation 7.10 do not equal the voltage differences that would be obtained if the bus voltage from the load flow were subtracted.

When the line a-b is removed, the current that had been flowing in it must now find alternate paths. This produces new voltage differences across the lines as indicated by equation 7.11 in which the Z-matrix elements are obtained from the matrix of Fig. 7.4 after the line a-b had been removed.

$$
\begin{array}{c} r\text{-}s \\ \\ u\text{-}v \end{array}
\left[\begin{array}{c} a \\ \\ Z' \\ \\ \end{array} \right]
\left[\begin{array}{c} \\ I_a \\ \\ \end{array} \right]
=
\left[\begin{array}{c} E'_r - E'_s \\ \\ E'_u - E'_v \end{array} \right]
\tag{7.11}
$$

The same current injection is being used but it causes a change in the voltage differences which are the result of the redistribution of current in the entire network.

$$
\left.
\begin{array}{l}
\Delta E_{rs} = (E'_r - E'_s) - (E_r - E_s) \\
\Delta E_{uv} = (E'_u - E'_v) - (E_u - E_v)
\end{array}
\right\}
\tag{7.12}
$$

The increments of current transferred to the lines are given by equation 7.13 in which the voltage differences are divided by the line impedances

$$
\left.
\begin{array}{l}
\Delta I_{rs} = \dfrac{\Delta E_{rs}}{Z_{\text{line}_{rs}}} = \dfrac{(E'_r - E'_s) - (E_r - E_s)}{Z_{\text{line}_{rs}}} \\[3mm]
\Delta I_{uv} = \dfrac{\Delta E_{uv}}{Z_{\text{line}_{uv}}} = \dfrac{(E'_u - E'_v) - (E_u - E_v)}{Z_{\text{line}_{uv}}}
\end{array}
\right\}
\tag{7.13}
$$

The $Z_{\text{line}_{rs}}$ and $Z_{\text{line}_{uv}}$ are line impedances obtained from the library record of the base case load flow.

The precontingency line flows can be computed from the voltages obtained from the library record

$$
\bar{I}_{rs} = \frac{\bar{E}_r - \bar{E}_s}{Z_{\text{line}_{rs}}}
\tag{7.14}
$$

in which the bar indicates base case power flow values. The current flow after the removal of line a-b is

$$
I'_{rs} = \bar{I}_{rs} + \Delta I_{rs}
\tag{7.15}
$$

The line flow power $S = P + jQ$ is obtained by multiplying the new current flow by the appropriate bus voltages obtained from the base case power flow.

A MULTIPLE CONTINGENCY

Assume that lines a-b, c-d, and e-f are to be opened simultaneously. The current vector must now satisfy three conditions. The voltage differences across the three lines must be made to be the same voltage differences that exist across these lines in the base case load flow.

Extract the matrix elements from the matrix that has been saved (Fig. 7.1) as required in equation 7.16.

$$
\begin{array}{c}
a \\ b \\ c \\ d \\ e \\ f
\end{array}
\begin{array}{cccccc}
a & b & c & d & e & f
\end{array}
\left[\quad Z \quad \right]
\begin{bmatrix} I_a \\ 0 \\ I_c \\ 0 \\ I_e \\ 0 \end{bmatrix}
=
\begin{bmatrix} E_a \\ E_b \\ E_c \\ E_d \\ E_e \\ E_f \end{bmatrix}
\tag{7.16}
$$

Three conditions must be met and three currents are to be determined that will satisfy these conditions. Note the current could be injected into any three buses of the system but again as a matter of convenience the buses selected are buses at the ends of the lines to be removed. The same condensation of matrix size is repeated by subtracting every second row from the row ahead in both the Z-matrix and the voltage vector.

$$
\begin{array}{c}
a\text{-}b \\ c\text{-}d \\ e\text{-}f
\end{array}
\begin{array}{cccccc}
a & b & c & d & e & f
\end{array}
\left[\quad Z \quad \right]
\begin{bmatrix} I_a \\ 0 \\ I_c \\ 0 \\ I_e \\ 0 \end{bmatrix}
=
\begin{bmatrix} E_a - E_b \\ E_c - E_d \\ E_e - E_f \end{bmatrix}
\tag{7.17}
$$

Omitting every second column in the Z-matrix and the zero current values affects a further condensation.

$$
\begin{array}{c}
\phantom{a\text{-}b} \\
a\text{-}b \\
c\text{-}d \\
e\text{-}f
\end{array}
\overset{\displaystyle a\,c\,e}{\left[\ \ Z\ \ \right]}
\begin{bmatrix} I_a \\ I_c \\ I_e \end{bmatrix}
=
\begin{bmatrix} E_a - E_b \\ E_c - E_d \\ E_e - E_f \end{bmatrix}
\tag{7.18}
$$

The voltage differences are known from the converged load flow case. The Z-matrix elements are known, and the currents must be determined. Inversion of the Z-matrix makes it possible to determine the required currents.

$$
\begin{bmatrix} I_a \\ I_c \\ I_e \end{bmatrix}
=
\begin{array}{c}
a\text{-}b \\
c\text{-}d \\
e\text{-}f
\end{array}
\overset{\displaystyle a\,c\,e}{\left[\ \ Z^{-1}\ \ \right]}
\begin{bmatrix} E_a - E_b \\ E_c - E_d \\ E_e - E_f \end{bmatrix}
\tag{7.19}
$$

It must be noted that the buses used for current injection must be independent buses. A single injection current can only satisfy one constraint. Therefore the injection points must be independent. With the three currents obtained from equation 7.19 it is necessary to determine the change in current that will occur in lines r-s and u-v when the three lines a-b, c-d, and e-f are opened simultaneously. The matrix of Fig. 7.5 is extracted from the matrix of Fig. 7.1.

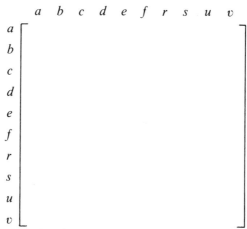

Fig. 7.5. Z-matrix before lines a-b, c-d, and e-f have been removed.

Using the building algorithm, three lines with impedances of opposite sign to that of the original data are added one at a time between buses a-b, c-d, and e-f to give the modified matrix of Fig. 7.6.

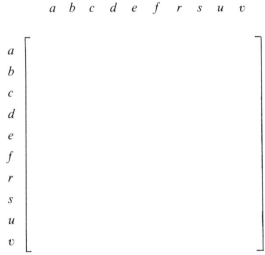

Fig. 7.6. Z'-matrix after lines a-b, c-d, and e-f have been removed.

By extracting the appropriate matrix elements from the matrix of Fig. 7.5, the voltage differences across the lines r-s and u-v produced by the three injection currents from equation 7.19 are determined by equation 7.20,

$$
\begin{array}{c} r\text{-}s \\ \\ u\text{-}v \end{array}
\begin{array}{c} a \quad b \quad c \quad d \quad e \quad f \\ \left[\qquad\qquad Z \qquad\qquad \right] \end{array}
\begin{bmatrix} I_a \\ 0 \\ I_c \\ 0 \\ I_e \\ 0 \end{bmatrix}
=
\begin{bmatrix} E_r - E_s \\ \\ E_u - E_v \end{bmatrix}
\tag{7.20}
$$

which can also be reduced to give

$$
\begin{array}{c} \\ r\text{-}s \\ \\ u\text{-}v \end{array}
\overset{a \quad c \quad e}{\left[\quad Z \quad \right]}
\left[\begin{array}{c} I_a \\ I_c \\ I_e \end{array} \right]
=
\left[\begin{array}{c} E_r - E_s \\ \\ E_u - E_v \end{array} \right]
\tag{7.21}
$$

A similar extraction of elements from the matrix of Fig. 7.6 gives the voltage differences after the line removals.

$$
\begin{array}{c} \\ r\text{-}s \\ \\ u\text{-}v \end{array}
\overset{a \quad c \quad e}{\left[\quad Z' \quad \right]}
\left[\begin{array}{c} I_a \\ I_c \\ I_e \end{array} \right]
=
\left[\begin{array}{c} E_r' - E_s' \\ E_u' - E_v' \end{array} \right]
\tag{7.22}
$$

The change in voltage differences reflects the change in distribution when the currents that formerly flowed from a to b, c to d, and e to f were forced to find alternate paths when those lines were removed.

The calculation of the increments of current follows exactly the method indicated for the single contingency.

$$
\Delta E_{rs} = (E_r' - E_s') - (E_r - E_s)
\tag{7.23}
$$

$$
\Delta I_{rs} = \frac{\Delta E_{rs}}{Z_{\text{line}_{rs}}} = \frac{(E_r' - E_s') - (E_r - E_s)}{Z_{\text{line}_{rs}}}
\tag{7.24}
$$

and

$$
I_{rs}' = \bar{I}_{rs} + \Delta I_{rs}
\tag{7.25}
$$

CONTINGENCY CAUSED BY A LINE ADDITION

A contingency need not necessarily be the loss of a transmission line. Closing a line that was open or building a new line can cause a redistribution of the flow in the lines of a system that might result in some lines becoming overloaded. The Z-matrix is not restricted to the investigation of contingencies caused by line openings. Consider the system shown in Fig. 7.7.

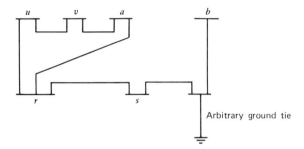

Fig. 7.7.

It is to be determined whether closing (or building) a line between buses a and b will produce overloads in lines sensitive to system changes. The matrix for the normal system (line open) has been built and is available (Fig. 7.1). The base case load flow, also with the line missing, is available in the computer library. Injection current I_a into bus a must be determined such that the voltage difference $E_a - E_b$ from matrix equation 7.26 is equal to the voltage difference obtained from the base case power flow.

$$
a\text{-}b\begin{array}{c} a \quad b \\ \left[\begin{array}{c} Z \end{array}\right] \end{array}\left[\begin{array}{c} I_a \\ \\ 0 \end{array}\right] = [E_a - E_b] \tag{7.26}
$$

Then the required current is

$$
I_a = \frac{E_a - E_b}{Z_{aa} - Z_{ba}} \tag{7.27}
$$

Injection of this current into bus a will produce the same voltage difference between buses a and b as exists in the base case load flow. The current flow from a to b will of course be zero, since the line is open and the line impedance, $Z_{\text{line}_{ab}}$ is therefore infinity. This is a slight variation in viewpoint from the previous one when current is injected into a bus to cause the base case current to flow over the line to be removed. There is complete agreement however between this and the previous work if the current injection is viewed as producing a voltage difference between two buses rather than a flow of current.

The portion of the Z-matrix corresponding to line a-b and the limiting lines r-s and u-v is extracted from the matrix of Fig. 7.1 that had been saved at the beginning of the series of contingency investigations (see Fig. 7.8).

$$
\begin{array}{c}
\begin{array}{cccccc} a & b & r & s & u & v \end{array} \\
\begin{array}{c} a \\ b \\ r \\ s \\ u \\ v \end{array}
\left[\phantom{\begin{array}{cccccc} a & b & r & s & u & v \\ & & & & & \\ & & & & & \\ & & & & & \\ & & & & & \\ & & & & & \end{array}} \right]
\end{array}
$$

Fig. 7.8. Z-matrix before adding the new line a-b.

The matrix is modified by adding the new line using the building algorithm to give the matrix of Fig. 7.9.

$$
\begin{array}{c}
\begin{array}{cccccc} a & b & r & s & u & v \end{array} \\
\begin{array}{c} a \\ b \\ r \\ s \\ u \\ v \end{array}
\left[\phantom{\begin{array}{cccccc} a & b & r & s & u & v \\ & & & & & \\ & & & & & \\ & & & & & \\ & & & & & \\ & & & & & \end{array}} \right]
\end{array}
$$

Fig. 7.9. Z'-matrix after adding the line a-b.

From the matrix of Fig. 7.8 the appropriate elements are selected for equation 7.28,

$$
\begin{array}{c}
\begin{array}{cc} a & b \end{array} \\
\begin{array}{c} r \\ s \\ u \\ v \end{array}
\left[Z \right]
\end{array}
\begin{bmatrix} I_a \\ 0 \end{bmatrix}
=
\begin{bmatrix} E_r \\ E_s \\ E_u \\ E_v \end{bmatrix}
\tag{7.28}
$$

which is reduced to equation 7.29 by subtracting rows of the Z-matrix and elements in the voltage vector. A column of the Z-matrix is omitted and the zero element of the current vector deleted.

$$\begin{matrix} r\text{-}s \\ u\text{-}v \end{matrix} \begin{bmatrix} & a & \\ & Z & \end{bmatrix} \begin{bmatrix} & \\ I_a & \\ & \end{bmatrix} = \begin{bmatrix} E_r - E_s \\ E_u - E_v \end{bmatrix} \tag{7.29}$$

From the matrix of Fig. 7.9 elements are extracted and reduced exactly as was done in arriving at equation 7.29 to give equation 7.30.

$$\begin{matrix} r\text{-}s \\ u\text{-}v \end{matrix} \begin{bmatrix} & a & \\ & Z' & \end{bmatrix} \begin{bmatrix} & \\ I_a & \\ & \end{bmatrix} = \begin{bmatrix} E_r' - E_s' \\ E_u' - E_v' \end{bmatrix} \tag{7.30}$$

From which it follows directly as before that

$$\Delta I_{rs} = \frac{(E_r' - E_s') - (E_r - E_s)}{Z_{line_{rs}}} \tag{7.31}$$

Thus contingencies caused by line addition can be treated by the method with equal facility to line removals.

INTERCHANGE CAPABILITY EVALUATION (SINGLE)

It may be desirable to determine the maximum amount of interchange between two neighboring companies that can be tolerated before the first of the limiting lines of the system will be loaded up to its thermal rating. The system may be in a line outage contingency condition. If there is a line outage contingency involved, the analysis is carried out exactly as has been shown. Then in addition to the injection currents shown in equation 7.18, the system is subjected to two other injection currents that simulate the shift in the interchange generation schedule. If no line outages are involved, omit the steps relating to line openings. The system being studied is shown in Fig. 7.10.

One of the transformers between a and b as well as one of the transformers between c and d are to be opened. These contingencies and a shift of generation will tend to overload lines r-s and u-v. The amount of generation input to the system at buses G_1 and G_2 are to be altered.

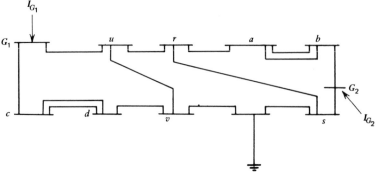

Fig. 7.10. Interchange evaluation.

The analysis is similar to the multiple contingency case except that the matrix that is extracted from the matrix of Fig. 7.1, which was saved at the beginning of the study, must now include the two generator axes (see Fig. 7.11).

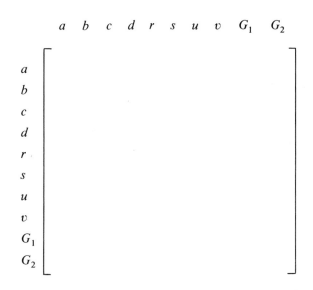

Fig. 7.11. Z-matrix before lines a-b and c-d have been removed.

The matrix is modified to correspond to the system with the two lines removed to give the matrix of Fig. 7.12.

$$
\begin{array}{c}
\begin{array}{cccccccccc} a & b & c & d & r & s & u & v & G_1 & G_2 \end{array} \\
\begin{array}{c} a \\ b \\ c \\ d \\ r \\ s \\ u \\ v \\ G_1 \\ G_2 \end{array}
\left[\right]
\end{array}
$$

Fig. 7.12. Z'-matrix after the lines a-b and c-d have been removed.

The determination of the currents I_a and I_c which will produce the required voltage drops across the line a-b and c-d follows exactly the technique of equation 7.16, 7.17, 7.18, and 7.19. The precontingency matrix equation is formed as shown in equation 7.32 from elements of the matrix of Fig. 7.11.

$$
\begin{array}{c}
\begin{array}{cccccc} a & b & c & d & G_1 & G_2 \end{array} \\
\begin{array}{c} r \\ s \\ u \\ v \end{array}
\left[\quad Z \quad \right]
\end{array}
\begin{bmatrix} I_a \\ 0 \\ I_c \\ 0 \\ 0 \\ 0 \end{bmatrix}
=
\begin{bmatrix} E_r \\ E_s \\ E_u \\ E_v \end{bmatrix}
\qquad (7.32)
$$

Note that at this time the currents I_{G_1} and I_{G_2} are both zero because the base case interchange is considered to be "normal." This equation is reduced to

$$
\begin{array}{c} \\ r\text{-}s \\ u\text{-}v \end{array}
\begin{array}{c} a \quad c \quad G_1 \quad G_2 \\ \left[\quad\quad Z \quad\quad \right] \end{array}
\begin{bmatrix} I_a \\ I_c \\ 0 \\ 0 \end{bmatrix}
=
\begin{bmatrix} E_r - E_s \\ E_u - E_v \end{bmatrix}
\tag{7.33}
$$

The equation representing the system after the double contingency, but before the interchange has been altered, is given by equation 7.34. The matrix elements are obtained from the matrix of Fig. 7.12.

$$
\begin{array}{c} \\ r\text{-}s \\ u\text{-}v \end{array}
\begin{array}{c} a \quad c \quad G_1 \quad G_2 \\ \left[\quad\quad Z' \quad\quad \right] \end{array}
\begin{bmatrix} I_a \\ I_c \\ 0 \\ 0 \end{bmatrix}
=
\begin{bmatrix} E'_r - E'_s \\ E'_u - E'_v \end{bmatrix}
\tag{7.34}
$$

A trial interchange of say ΔP pu power is converted to injection current by use of equation 7.35.

$$
I_{G_1} = \frac{\Delta P + jQ}{E^*_{G_1}}
\tag{7.35}
$$

The $E^*_{G_1}$ is obtained from the base case power flow. The current I_{G_1} is used in the current vector to represent a change of generation being supplied to the system at bus G_1. A simultaneous counter adjustment at bus G_2 simulates a revision in the interchange schedule, an increase in generation at bus G_1, and a decrease at G_2.

Equation 7.36 represents the system after the lines have been removed and the shift of ΔP pu power has been made in the interchange schedule.

$$
\begin{array}{c} \\ r\text{-}s \\ u\text{-}v \end{array}
\begin{array}{c} a \quad c \quad G_1 \quad G_2 \\ \left[\quad\quad Z' \quad\quad \right] \end{array}
\begin{bmatrix} I_a \\ I_c \\ I_{G_1} \\ I_{G_2} \end{bmatrix}
=
\begin{bmatrix} E''_r - E''_s \\ E''_u - E''_v \end{bmatrix}
\tag{7.36}
$$

Three current values can be computed for the flow between buses r and s.

The original current flow is computed using $\overline{E}_r$ and $\overline{E}_s$ from the base case power flow.

$$I_{rs} = \frac{\overline{E}_r - \overline{E}_s}{Z_{line_{rs}}} \tag{7.37}$$

The flow after the line removals is computed using the voltage values obtained from equation 7.34.

$$I'_{rs} = I_{rs} + \Delta I_{rs} = I_{rs} + \frac{(E'_r - E'_s) - (\overline{E}_r - \overline{E}_s)}{Z_{line_{rs}}} \tag{7.38}$$

The flow in the line, after the lines have been removed and the interchange schedule has been modified by an increment (or decrement) of ΔP MW from the interchange existing in the base case power flow, is obtained from equation 7.39 using voltages from equation 7.36.

$$I''_{rs} = I_{rs} + \Delta I'_{rs} = I_{rs} + \frac{(E''_r - E''_s) - (\overline{E}_r - \overline{E}_s)}{Z_{line_{rs}}} \tag{7.39}$$

Using these current values and the bus voltage $\overline{E}_r$ from the base case load flow the complex power flowing into the line from bus r can be computed.

$$P_{rs} + jQ_{rs} = I^*_{rs} \overline{E}_r \tag{7.40}$$

$$P'_{rs} + jQ'_{rs} = I'^*_{rs} \overline{E}_r \tag{7.41}$$

$$P''_{rs} + jQ''_{rs} = I''^*_{rs} \overline{E}_r \tag{7.42}$$

MAXIMUM INTERCHANGE DETERMINATION

The line flow before and the line flow after the revision in the scheduled amount of interchange establishes two relationships: line MW versus line $MVAR$ and line MW versus interchange MW (see Figs. 7.13 and 7.14).

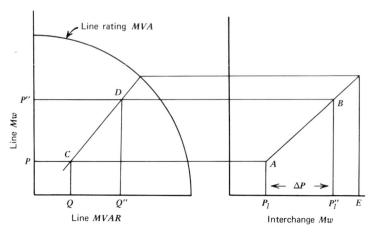

Fig. 7.13.   Extrapolation to determine maximum interchange permitted.

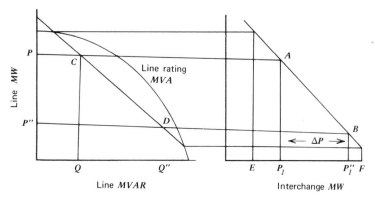

Fig. 7.14.   Interchange restricted between two limits.

These relationships can be used to extrapolate or interpolate the permissable amount of interchange before the line will be operating at its *MVA* limit. This calculation is made for all the limiting lines. The smallest interchange obtained for the several lines tested is the maximum permissable interchange that the system can tolerate.

Great care must be used in interpreting the results of these investigations. The logic must be carefully considered because of the many cases that can result. The figures illustrate two of the many combinations that can occur. In each figure the following convention is adopted:

A = line MW versus interchange MW in the base case.

B = line MW versus interchange MW when the interchange has been increased by ΔP pu MW.

C = line MW versus line $MVAR$ in the base case.

D = line MW versus line $MVAR$ in the increased interchange case.

E and F = permissible interchange which loads the line to its MVA rating.

The single interchange capability evaluation technique described in the previous topic is helpful in solving planning and operating problems involving interchange with an individual neighbor but very few, if any, companies have only one interconnection. A more general problem is what the optimum interchange schedule between several companies is, such that, the next contingency (of several possible contingencies) will not overload any of the limiting lines of the interconnected system. The first step in solving this problem was made by using the output of the single interchange program as input to a linear programming code.*

By studying the results of a great number of contingency cases it can be determined that the limiting lines r-s, u-v, and w-x are the most sensitive to the loss of critical lines a-b, c-d, and e-f, respectively. This information is then prepared as input to the linear programming code, which determines the optimum interchange schedule, such that, the loss of line a-b will not overload the line r-s (which is the most sensitive to the removal of a-b); the loss of c-d will not overload u-v, and the loss of e-f will not overload w-x.

The preparation of the data necessary for the linear programming program requires the determination of distribution factors computed from the output of the single interchange program. A distribution factor is the fraction of the current of a line, which is transferred to a second line when the first line is opened. Since distribution factors can be determined directly from Z-matrix elements [2, 3, 15], it was natural that a Z-matrix method would be evolved which would determine the optimum interchange without the necessity of resorting to a linear programming code [14].

Seldom, if ever, is the base case power flow that is available in the library the actual system that is to be studied. Lines that were in the base

---

* Linear programming, a facit of mathematical programming, does not mean writing computer programs to solve mathematical problems but is, rather, a mathematical procedure for finding a maximum or minimum of a function subject to inequality constraints. The name was invented in 1947 when computer programming was called coding. The subject of linear programming is discussed as one of the topics of Chapter 9.

case are open for maintenance, construction work, or some other contingency. Other lines may have been added to the base case system, or lines that were opened now have been reclosed. It is this modified system that must be studied to optimize the interchange, subject to the constraint that the next contingency will not produce overloaded lines. The steps that must be taken in the analysis are as follows.

1. By means of the contingency evaluation program, compute the change in line flows in the critical and limiting lines when the base case system is converted to the new system condition by the addition and removal of lines.

2. Using the same technique remove each critical line, one at a time, and determine the additional amount of flow in the limiting line which is most sensitive to each particular line removal contingency condition.

3. The remaining capability of each limiting line after absorbing the incremental flows of step 1 and step 2 is determined.

4. The interchange schedule is adjusted to load each limiting line to its rated capability when the single contingency, to which it is most sensitive, occurs.

The matrix that was saved in Fig. 7.1 is enlarged to include the axes corresponding to the line changes necessary to convert the base case power flow to the system to be studied (see Fig. 7.15).

It should be born in mind that in actual practice, the axes of the matrix are not segregated as shown in Fig. 7.15, but rather, the axes occur in a random order during the matrix building routine. A catalogue is kept of the location in memory of each row of the matrix, so that any element may be extracted as needed in the mathematical procedure. The matrix is shown with columns arranged by categories as an aid to explaining the method. From the large matrix of Fig. 7.15 are extracted the axes of interest in the particular case to be studied (see Fig. 7.16).

The current in all the lines of the study (line change, critical lines, and limiting lines) are computed by means of equation 7.43, using the base case power flow voltages. Line a-b is used as an example.

$$\bar{I}_{ab} = \frac{\bar{E}_a - \bar{E}_b}{Z_{line_{ab}}} \qquad (7.43)$$

The overbar indicates the voltage values are obtained from the base case power flow.

	Line changes	Critical lines	Limiting lines	Generators
	$i\ j\ k\ l\ m\ n \cdots$	$a\ b\ c\ d\ e\ f \cdots$	$r\ s\ u\ v\ w\ x \cdots$	$A\ B\ C \cdots$

$$
\begin{array}{c}
i \\ j \\ k \\ l \\ m \\ n \\ \vdots \\ a \\ b \\ c \\ d \\ e \\ f \\ \vdots \\ r \\ s \\ u \\ v \\ w \\ x \\ \vdots \\ A \\ B \\ C \\ \vdots
\end{array}
\left[\right]
$$

Fig. 7.15. Z-matrix retained. All other axes have been discarded.

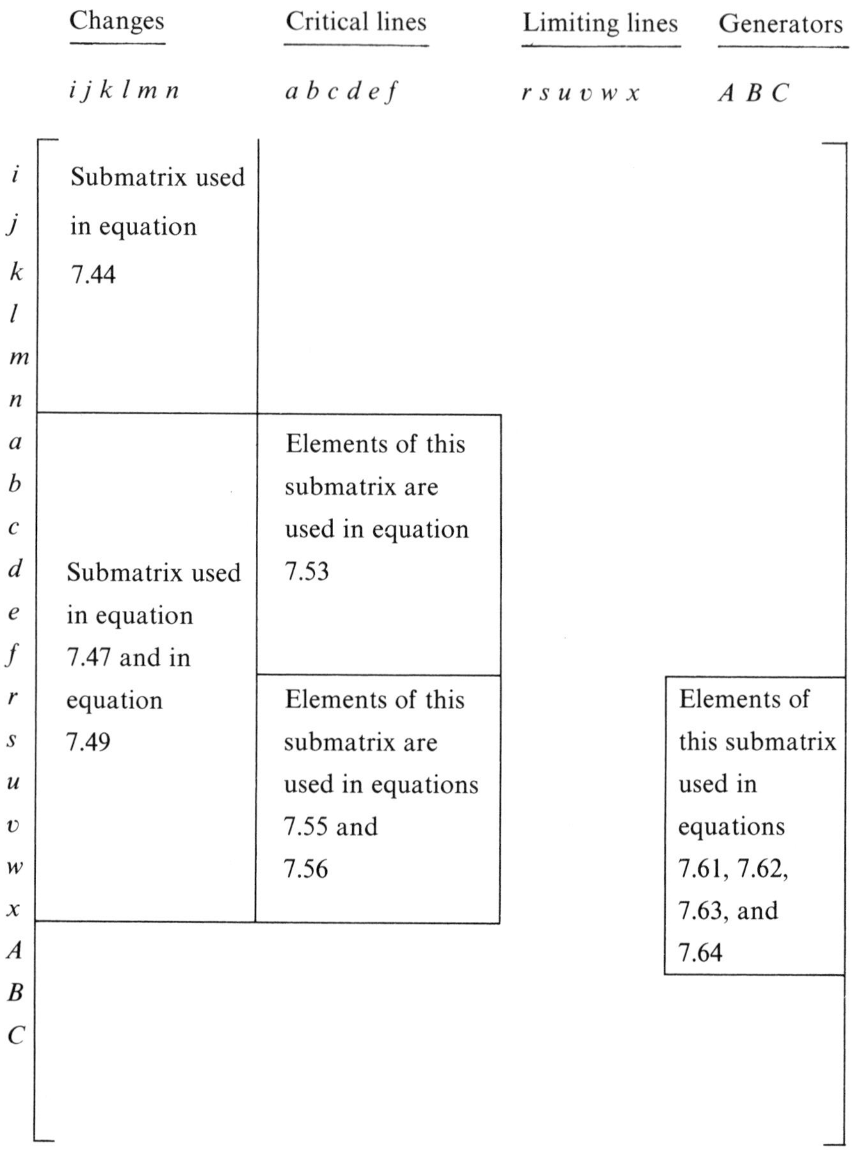

Changes	Critical lines	Limiting lines	Generators
$i\,j\,k\,l\,m\,n$	$a\,b\,c\,d\,e\,f$	$r\,s\,u\,v\,w\,x$	$A\,B\,C$

	Changes	Critical lines	Limiting lines	Generators
i	Submatrix used in equation 7.44			
j				
k				
l				
m				
n				
a		Elements of this submatrix are used in equation 7.53		
b				
c				
d	Submatrix used in equation 7.47 and in equation 7.49			
e				
f				
r		Elements of this submatrix are used in equations 7.55 and 7.56		Elements of this submatrix used in equations 7.61, 7.62, 7.63, and 7.64
s				
u				
v				
w				
x				
A				
B				
C				

Fig. 7.16. The parts of the Z matrix used in a particular study.

The Z-matrix of equation 7.44 is extracted from the matrix of Fig. 7.16. The current injections are to be determined such that the voltage difference across the lines to be removed will be the same voltage difference that occurs across the lines in the power flow base case.

$$
\begin{array}{c}
\\
i \\
j \\
k \\
l \\
m \\
n
\end{array}
\begin{array}{c}
i \; j \; k \; l \; m \; n \\
\left[\qquad\qquad\qquad \right]
\end{array}
\left[\begin{array}{c} I_i \\ 0 \\ I_k \\ 0 \\ I_m \\ 0 \end{array} \right]
=
\left[\begin{array}{c} E_i \\ E_j \\ E_k \\ E_l \\ E_m \\ E_n \end{array} \right]
\tag{7.44}
$$

with Z in the matrix.

This equation reduces to equation 7.45 by the method described earlier in this chapter.

$$
\begin{array}{c}
\\
i\text{-}j \\
k\text{-}l \\
m\text{-}n
\end{array}
\begin{array}{c}
i \;\; k \;\; m \\
\left[\quad Z \quad \right]
\end{array}
\left[\begin{array}{c} I_i \\ I_k \\ I_m \end{array} \right]
=
\left[\begin{array}{c} E_i - E_j \\ E_k - E_l \\ E_m - E_n \end{array} \right]
\tag{7.45}
$$

The voltage differences are obtained from the power flow base case. The current injections into buses i, k, and m which will produce these voltage differences must be determined by inversion of the Z-matrix as shown in equation 7.46.

$$
\left[\begin{array}{c} I_i \\ I_k \\ I_m \end{array} \right]
=
\begin{array}{c}
\\
i\text{-}j \\
k\text{-}l \\
m\text{-}n
\end{array}
\begin{array}{c}
i \;\; k \;\; m \\
\left[\quad Z^{-1} \quad \right]
\end{array}
\left[\begin{array}{c} E_i - E_j \\ E_k - E_l \\ E_m - E_n \end{array} \right]
\tag{7.46}
$$

The current injections that have been determined cause the proper voltage difference to occur across the lines to be removed (or added). Thus the lines will be carrying their proper power flow current when the lines i-j, *k-l*, and *m-n* are removed; the currents that they have been carrying must be redistributed throughout the network. The current in the lines of interest in this study, *a-b*, *c-d*, *e-f* (critical lines), and *r-s*, *u-v*, and *w-x* (limiting lines), must be modified to reflect this redistribution of current. The matrix of equation 7.47 is extracted from the matrix of Fig. 7.16.

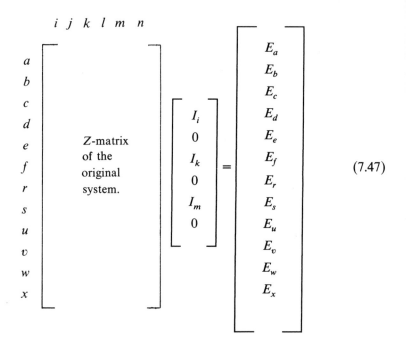

$$
\begin{array}{c} a \\ b \\ c \\ d \\ e \\ f \\ r \\ s \\ u \\ v \\ w \\ x \end{array}
\begin{matrix} i & j & k & l & m & n \end{matrix}
\left[\quad Z\text{-matrix of the original system.} \quad \right]
\begin{bmatrix} I_i \\ 0 \\ I_k \\ 0 \\ I_m \\ 0 \end{bmatrix}
=
\begin{bmatrix} E_a \\ E_b \\ E_c \\ E_d \\ E_e \\ E_f \\ E_r \\ E_s \\ E_u \\ E_v \\ E_w \\ E_x \end{bmatrix}
\tag{7.47}
$$

This matrix equation is reduced as before to give equation 7.48.

$$
\begin{array}{c} a\text{-}b \\ c\text{-}d \\ e\text{-}f \\ r\text{-}s \\ u\text{-}v \\ w\text{-}x \end{array}
\begin{matrix} i & k & m \end{matrix}
\left[\quad Z \quad \right]
\begin{bmatrix} I_i \\ I_k \\ I_m \end{bmatrix}
=
\begin{bmatrix} E_a - E_b \\ E_c - E_d \\ E_e - E_f \\ E_r - E_s \\ E_u - E_v \\ E_w - E_x \end{bmatrix}
\tag{7.48}
$$

The matrix of Fig. 7.16 is modified for the removal (or addition) of lines *i-j*, *k-l*, and *m-n* by the matrix building algorithm. In all subsequent equations a matrix designated by Z' will indicate a matrix formed from elements of the matrix of Fig. 7.16 after these line changes have been made (see Fig. 7.17).

$$
\begin{array}{c}
\quad\quad i \quad k \quad m \\
\begin{array}{c}
a\text{-}b \\
c\text{-}d \\
e\text{-}f \\
r\text{-}s \\
u\text{-}v \\
w\text{-}x
\end{array}
\left[\begin{array}{ccc}
\quad & \quad & \quad \\
\quad & \quad & \quad \\
\quad & \quad & \quad \\
\quad & \quad & \quad \\
\quad & \quad & \quad \\
\quad & \quad & \quad
\end{array}\right]
\end{array}
$$

Fig. 7.17. Z'-matrix is the matrix of Fig. 7.16 modified for line changes *i-j*, *k-l*, and *m-n*.

From the modified matrix of Fig. 7.17 are extracted the necessary elements for the matrix equation 7.49. In this equation the subtraction of rows and the deletion of columns have been performed.

$$
\begin{array}{c}
\quad\quad i \quad k \quad m \\
\begin{array}{c}
a\text{-}b \\
c\text{-}d \\
e\text{-}f \\
r\text{-}s \\
u\text{-}v \\
w\text{-}x
\end{array}
\left[\begin{array}{c}
Z'\text{-matrix} \\
\text{after line} \\
\text{changes} \\
i\text{-}j,\ k\text{-}l, \\
\text{and } m\text{-}n \text{ have} \\
\text{been completed.}
\end{array}\right]
\left[\begin{array}{c}
I_i \\
\\
I_k \\
\\
I_m
\end{array}\right]
=
\left[\begin{array}{c}
E'_a - E'_b \\
E'_c - E'_d \\
E'_e - E'_f \\
E'_r - E'_s \\
E'_u - E'_v \\
E'_w - E'_x
\end{array}\right]
\end{array}
\qquad (7.49)
$$

The increment of current added to each of the critical and limiting lines when the system is modified to the system condition to be studied, from the base case condition, is computed using line *a-b* as an example in equation 7.50. The voltage differences used in this equation are obtained from equations 7.48 and 7.49.

$$
\Delta I_{a\text{-}b} = \frac{(E'_a - E'_b) - (E_a - E_b)}{Z_{\text{line}_{a\text{-}b}}}
\qquad (7.50)
$$

The original current flow from the base case power flow condition determined by equation 7.43 is augmented by this increment of current to give the current flow in the line of the modified base case (see equation 7.51).

$$I'_{a\text{-}b} = \bar{I}_{a\text{-}b} + \Delta I_{a\text{-}b} = \frac{\bar{E}_a - \bar{E}_b + (E'_a - E'_b) - (E_a - E_b)}{Z_{line_{a\text{-}b}}} \tag{7.51}$$

This calculation is repeated for all critical and limiting lines. The voltage difference across the line for this modified base condition and the new current flow in the line have been determined. As a matter of convenience the voltage difference can be expressed as

$$E''_a - E''_b = I'_{a\text{-}b} Z_{line_{a\text{-}b}} \tag{7.52}$$

Only single contingencies are now being considered. The current injection into bus a to produce the difference of voltage given by equation 7.52 is determined by using elements from the Z'-matrix of Fig. 7.17.

$$\begin{matrix} & a \\ a\text{-}b[& Z' \,] \end{matrix} [I_a] = E''_a - E''_b \tag{7.53}$$

$$I_a = \frac{E''_a - E''_b}{Z'_{a\text{-}a} - Z'_{b\text{-}a}} \tag{7.54}$$

It had previously been established by analysis of many contingency cases that line r-s is the line that is the most susceptible to overloading when line a-b is opened. To determine the additional increment of current transferred to this line when line a-b is opened, extract the matrix of Fig. 7.18 from the matrix Z' of Fig. 7.17. Extra axes ABC are included in the matrix at this time in anticipation of their use at a later time (see equation 7.61).

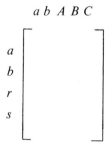

Fig. 7.18. Z'-matrix elements.

Remove the line a-b by means of the building algorithm and obtain the matrix of Fig. 7.19.

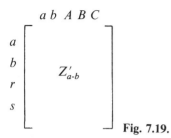

Fig. 7.19.

Note: The matrix designation $Z'_{a\text{-}b}$ was introduced in Ref. 14. It indicates that the modification of the original matrix Z for the removal of lines i-j, k-l, and m-n to give the matrix Z' has now been modified for the removal of line a-b.

Elements from the matrices of Figs. 7.18 and 7.19 are extracted to form the matrices of equations 7.55 and 7.56. The current was determined by equation 7.54.

$$r\text{-}s \; [\quad Z' \quad] \, [I_a] = [E'_r - E'_s] \tag{7.55}$$

$$r\text{-}s \; [Z'_{a\text{-}b}] \, [I_a] = [E''_r - E''_s] \tag{7.56}$$

Opening line a-b causes line r-s to pick up an additional increment of current that resulted from the change in voltage difference across the line r-s as shown in equation 7.57.

$$\Delta I'_{r\text{-}s} = \frac{(E''_r - E''_s) - (E'_r - E'_s)}{Z_{\text{line}_{r\text{-}s}}} \tag{7.57}$$

The total current flowing in the line r-s after line a-b has been opened is the original line current, plus the increment produced when the lines i-j, k-l, and m-n were opened, plus the increment attributable to the opening of line a-b (see equation 7.58).

$$I''_{r\text{-}s} = I'_{r\text{-}s} + \Delta I'_{r\text{-}s} = I_{r\text{-}s} + \Delta I_{r\text{-}s} + \Delta I'_{r\text{-}s} \tag{7.58}$$

The difference between this current and the current rating of the line is the capability of the line available for adjustments in the interchange schedule.

$$\Delta I''_{r\text{-}s} = I^{\text{rating}}_{r\text{-}s} - I''_{r\text{-}s} \tag{7.59}$$

This increment of current flowing in the line would produce an increment of voltage across the line.

$$(E'''_r - E'''_s) = \Delta I''_{r\text{-}s} Z_{\text{line}_{r\text{-}s}} \tag{7.60}$$

From the matrix of Fig. 7.19 select the required elements for equation 7.61.

$$
\begin{array}{c}
\\
r\\
\\
s
\end{array}
\begin{array}{c}
A\ B\ C\\
\left[\ Z'_{a\text{-}b}\ \right]
\end{array}
\left[\begin{array}{c} I_A \\ I_B \\ I_C \end{array}\right]
=
\left[\begin{array}{c} E'''_r \\ \\ E'''_s \end{array}\right]
\tag{7.61}
$$

The current to be injected into the generator buses that will simulate interchange must be determined subject to the constraint that the available voltage $(E'''_r - E'''_s)$ from equation 7.60 will be produced. Equation 7.61 is simplified to give this voltage difference in equation 7.62.

$$
r\text{-}s\
\begin{array}{c}
A\ B\ C\\
\left[\ Z'_{a\text{-}b}\ \right]
\end{array}
\left[\begin{array}{c} I_A \\ I_B \\ I_C \end{array}\right]
= [E'''_r - E'''_s]
\tag{7.62}
$$

In exactly the same manner the removal of the line $c\text{-}d$ must cause the associated limiting line $u\text{-}v$ to become loaded to its rated capacity. Beginning with equation 7.52, a, b, r, and s are replaced with c, d, u, and v, respectively. The resulting equation that must be satisfied is

$$
u\text{-}v\
\begin{array}{c}
A\ B\ C\\
\left[\ Z'_{c\text{-}d}\ \right]
\end{array}
\left[\begin{array}{c} I_A \\ I_B \\ I_C \end{array}\right]
= [E'''_u - E'''_v]
\tag{7.63}
$$

In the same manner, the equation is obtained for the effect of line e-f on the line w-x.

$$w\text{-}x \quad \begin{bmatrix} A & B & C \\ & Z'_{e\text{-}f} & \end{bmatrix} \begin{bmatrix} I_A \\ I_B \\ I_C \end{bmatrix} = [\,E'''_w - E'''_x\,] \qquad (7.64)$$

The three equations 7.62, 7.63, and 7.64 can be combined in a single matrix equation.

$$\begin{matrix} r\text{-}s \\ u\text{-}v \\ w\text{-}x \end{matrix} \begin{bmatrix} A & B & C \\ -Z'_{a\text{-}b} \rightarrow \\ -Z'_{c\text{-}d} \rightarrow \\ -Z'_{e\text{-}f} \rightarrow \end{bmatrix} \begin{bmatrix} I_A \\ I_B \\ I_C \end{bmatrix} = \begin{bmatrix} E'''_r - E'''_s \\ E'''_u - E'''_v \\ E'''_w - E'''_x \end{bmatrix} \qquad (7.65)$$

The fact that the three rows of the Z-matrix of equation 7.65 come from three different modifications of the Z'-matrix of Fig. 7.17 does not affect the validity of the equation. The voltage differences have been computed by equation 7.60. It remains to evaluate the current injections. By matrix inversion, equation 7.66 is obtained.

$$\begin{bmatrix} I_A \\ I_B \\ I_C \end{bmatrix} = \begin{matrix} r\text{-}s \\ u\text{-}v \\ w\text{-}x \end{matrix} \begin{bmatrix} A & B & C \\ -Z'_{a\text{-}b} \rightarrow \\ -Z'_{c\text{-}d} \rightarrow \\ -Z'_{e\text{-}f} \rightarrow \end{bmatrix}^{-1} \begin{bmatrix} E'''_r - E'''_s \\ E'''_u - E'''_v \\ E'''_w - E'''_x \end{bmatrix} \qquad (7.66)$$

These injection currents can be converted to generation megawatts by multiplication by the appropriate bus voltage from the base case load flow.

$$P_A + jQ_A = E_A I_A^* \qquad (7.67)$$

This is the incremental interchange generation injected into the bus over and above the amount that was occurring in the base case.

The method can handle special cases. For example, if a critical line a-b

influences two (or more) limiting lines, r-s and u-v, equation 7.65 becomes.

$$
\begin{array}{c}
\quad\quad A\ B\ C \\
\begin{array}{c} r\text{-}s \\ u\text{-}v \\ w\text{-}x \end{array}
\left[\begin{array}{c} -Z'_{a\text{-}b} \longrightarrow \\ -Z'_{a\text{-}b} \longrightarrow \\ -Z'_{c\text{-}d} \longrightarrow \end{array} \right]
\left[\begin{array}{c} I_A \\ I_B \\ I_C \end{array} \right]
=
\left[\begin{array}{c} E'''_r - E'''_s \\ E'''_u - E'''_v \\ E'''_w - E'''_x \end{array} \right]
\end{array}
\qquad (7.68)
$$

The Z-matrix is nonsingular since, it represents portions of different rows of the matrix of Fig. 7.17 corresponding to lines u-v and r-s. The matrix can therefore be inverted and the injection currents determined.

If the outage of two lines affect the same line equation 7.65 is written

$$
\begin{array}{c}
\quad\quad A\ B\ C \\
\begin{array}{c} r\text{-}s \\ r\text{-}s \\ w\text{-}x \end{array}
\left[\begin{array}{c} -Z'_{a\text{-}b} \longrightarrow \\ -Z'_{c\text{-}d} \longrightarrow \\ -Z'_{e\text{-}f} \longrightarrow \end{array} \right]
\left[\begin{array}{c} I_A \\ I_B \\ I_C \end{array} \right]
=
\left[\begin{array}{c} E'''_r - E'''_s \\ E'''_r - E'''_s \\ E'''_w - E'''_x \end{array} \right]
\end{array}
\qquad (7.69)
$$

This equation can also be solved for the current injections.

The formulation of the Z-matrix algorithm that has been described is by no means the only formulation of the technique. It was however the first implementation of the method and has been used extensively [5,6]. The use of a single ground tie of $1.0 + j1.0$ has appeared to be quite arbitrary and open to question (see discussions of Ref. 5) but the success of the program recommends this simplification.

It must be recognized that the Z-matrix method is an approximate method. Extensive testing on cases for a system of 2500 buses and 4500 lines in which a retained matrix of 150 by 150 from which hundreds of contingencies were studied gave results of within 2% [6]. Certain basic assumptions have been made, for example, that line outages do not change the total system load. Care must be taken that engineering judgement is used, and the algorithm is not required to attempt the impossible. For example, the line flows predicted by this method for a system that has been made steady-state unstable by a severe contingency will certainly not be valid. In that case, a load flow program would diverge, and the real system would fall apart, but this method cannot detect the difficulty.

To form the matrix and compute the first contingency requires about the same time required to do a load flow case by the Newton-Raphson method. The method has in this one case computed the equivalent of three load flow cases; the system under contingency conditions, under contingency conditions plus the increment of interchange, and extrapolated to the case of maximum allowable interchange condition. After the matrix and first contingency condition have been calculated, additional contingency and interchange cases are completed at 500 times the speed of the Newton-Raphson load flow program. It is seen that it is economic to stack hundreds of cases to be run concurrently. If only a single case or a few cases are required, it is probably more economic to compute contingencies by one of the fast Newton-Raphson methods. Both methods should be implemented but the tendency is certainly toward a greater and greater number of cases in each expansion study [6].

THE NEWTON-RAPHSON POWER FLOW

The original formulation of the power flow algorithm proposed by Ward and Hale [13] used a modified Newton-Raphson method in which the voltage adjustments made during the iterative process were determined by solving the set of simultaneous linear equations 7.70. All off-diagonal elements in the Jacobian matrix are equal to zero.

$$
\begin{bmatrix} \Delta P_k \\ \Delta Q_k \end{bmatrix} = \begin{bmatrix} \dfrac{\partial P_k}{\partial e_k} & \dfrac{\partial P_k}{\partial f_k} \\ \dfrac{\partial Q_k}{\partial e_k} & \dfrac{\partial Q_k}{\partial f_k} \end{bmatrix} \begin{bmatrix} \Delta e_k \\ \Delta f_k \end{bmatrix} \tag{7.70}
$$

In each iteration the adjustment to be made in the rectangular components of the bus voltages Δe_k and Δf_k that will ultimately reduce the mismatch of watts and vars to zero at every bus takes into account only the rate of change of the power at the bus with respect to its own voltage. The voltages at all other buses are assumed to be constant, and thus the rates of change of the power at a bus with repsect to the voltages of other buses are zero. The method was not used in the production programs that were implemented immediately after publication of the paper, because of the greater ease of programming the Gauss-Seidel algorithm [18].

In 1961 Van Ness [19] described a power flow method using the full Newton-Raphson representation in which the rate of change of the complex power delivered to a bus took into account the rate of change of the power with respect to the voltage magnitude and the angle of not only its own voltage but also the voltages and angles of all of its immediate neighbors. The equations which must be solved for the adjustments to be made in voltage magnitudes and angles to reduce the power mismatch at every bus are given by equation 7.71.

$$
\begin{bmatrix} \left(P_{k(\text{scheduled})} - P_{k(\text{computed})}\right) \\ \left(Q_{k(\text{scheduled})} - Q_{k(\text{computed})}\right) \end{bmatrix} = \begin{bmatrix} \Delta P_k \\ \Delta Q_k \end{bmatrix} = \begin{bmatrix} \dfrac{\partial P_k}{\partial \delta_m} & \dfrac{\partial P_k}{\partial E_m} \\ \dfrac{\partial Q_k}{\partial \delta_m} & \dfrac{\partial Q_k}{\partial E_m} \end{bmatrix} = \begin{bmatrix} \Delta \delta_m \\ \Delta E_m \end{bmatrix}
$$

$$(7.71)$$

These equations may be written more simply, using the notation of Van Ness.

$$
\begin{bmatrix} \Delta P_k \\ \Delta Q_k \end{bmatrix} = \begin{bmatrix} H_{km} & N_{km} \\ J_{km} & L_{km} \end{bmatrix} \begin{bmatrix} \Delta \delta_m \\ \dfrac{\Delta E_m}{E_m} \end{bmatrix} \qquad (7.72)
$$

The increment ΔE_m in equation 7.71 is divided by E_m to simplify computing the Jacobian matrix elements N_{km} and L_{km} as indicated in equations 6.26 and 6.27.

By optimizing the order in which the bus equations are entered in the matrix equation 7.72 and thus exploiting the sparcity of the matrix, Tinney [20,21] produced an algorithm that has become a standard in the power industry because of the speed, accuracy, and the ability to solve power flow cases that could not be solved by other methods.

REVIEW OF SOME FUNDAMENTALS

The power delivered to a bus depends to a greater extent on its own voltage than on the voltage of the neighboring buses because the diagonal element of the nodal admittance matrix is dominant (see equation 6.5). This dominance is true, in general, if there are no negative impedance branches connected to the bus. For this reason the simplified Jacobian matrix of Ward and Hale does succeed in solving power problems, but would of course fail to give solutions to some problems that can be solved by the full Jacobian matrix method. It is of interest that this simplified technique can solve a broad category of problems; however, the rate of convergence is comparitively poor and therefore would not be used for a fast method of evaluating contingencies.

A characteristic of electrical power transmission systems is the high reactance to resistance ratio of transmission lines and transformers. For this reason the watts transmitted from one bus to another over a transmission line depends primarily on the angular difference between the voltages of the two buses at the ends of the line and only to a very minor degree on the difference in the two voltage magnitudes. The vars, on the other hand, are strongly influenced by the difference in voltage magnitudes and only to a minor extent on the angular difference of the voltages of the two buses. In Fig. 7.20 two voltages of equal magnitude with an angular difference of δ are applied to the ends of a transmission line. The voltage difference between the two ends of the line is ΔE.

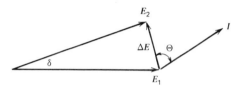

Fig. 7.20. Power flow depends on angular difference.

The current in the line I in Fig. 7.20 will lag the voltage difference ΔE by an angle θ which is nearly 90° because of the high X to R ratio of the line. For stability reasons the angle δ is usually small; therefore, the angle between I and E_1 (or E_2) is small and the watt component of the current is large compared to the reactive component.

In Fig. 7.21 two voltages that are applied to the ends of a line are in phase with each other but are of unequal magnitudes, thus producing a voltage difference ΔE across the line. The current in the line lags the voltage difference (and the bus voltages) by nearly 90°. The power transfer will be largely reactive. Since in a power system the voltage magnitudes are constrained to rather narrow limits, the ability to transmit vars is limited compared to the ability to transmit watts.

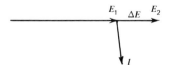

Fig. 7.21. Var flow depends on difference of voltage magnitudes.

Carpentier in 1963, recognizing this comparative independence of real power to changes of voltage magnitude and reactive power to changes in angle [22], suggested that N_{km} and J_{km} in equation 7.73 be set equal to zero; thus the watt and var equations would be decoupled as shown in equations 7.74 and 7.75.

$$[\Delta P_k] = [H_{km}][\Delta \delta_m] \tag{7.74}$$

$$[\Delta Q_k] = [L_{km}][\Delta E_m] \tag{7.75}$$

This decoupling would decrease the time required per iteration, since there are less calculations to made and the decoupling would reduce the memory requirement for storage of the program. The suggested technique attracted very little attention at that time because the money being spent on load flow calculations had not as yet increased to the point that it required immediate attention.

In the last half of the 1960s the rapid growth in the number of high voltage interconnections enlarged the area of influence in system studies far beyond company boundaries into neighboring systems. This required more and more buses to be used for a proper representation, with consequently longer running time per case, and a great increase in the number of contingency conditions that require investigation. With this incentive, the method suggested by Carpentier was reinvestigated.

The methods [7–9] that evolved not only decouple the real and reactive power equations 7.74 and 7.75 to achieve greater speed but also use additional clever techniques devised to simulate the outage of branches. These modified Newton-Raphson methods are undoubtedly more accurate than the Z-matrix method described earlier in this chapter but they do not achieve the great speed of that method.

THE DECOUPLING OF THE NEWTON-RAPHSON EQUATIONS

The complex node current I_k can be expressed in terms of the complex node to ground voltages of the buses $\bar{E}_k$ and the admittances of the lines connected to bus k in the network as given in equation 7.76.

$$\bar{I}_k = \sum_{m=1}^{N} \bar{Y}_{km} \bar{E}_{km} \tag{7.76}$$

The complex power being injected into bus k in terms of the node current and voltage is expressed by equation 7.77.

$$S_k = P_k + jQ_k = \bar{E}_k \bar{I}_k^* = \bar{E}_k \sum_{m=1}^{N} \bar{Y}_{km}^* \bar{E}_m^* \tag{7.77}$$

To simplify the explanation of decoupling the real and imaginary parts of this equation, it is rewritten with the voltages in polar form and the admittances in rectangular form in equation 7.78, in which δ_k and δ_m are the angular position of the bus voltages E_k and E_m with respect to a reference voltage vector.

$$P_k + jQ_k = \sum_{m=1}^{N} E_k \epsilon^{j\delta_k} E_m \epsilon^{-j\delta_m} (G_{km} - jB_{km})$$

$$= \Sigma E_k E_m \epsilon^{j(\delta_k - \delta_m)} (G_{km} - jB_{km}) \tag{7.78}$$

The term in the summation for $m = k$ can be removed from under the summation sign if the range of the index is properly adjusted to take this into account. Let α_k be defined as the set of branches connected to bus k. Then adopting the notation $m \in \alpha_k$ to mean that m is a member of the set of buses connected to bus k by a branch of the set α_k, the equation may be rewritten as shown in equation 7.79. Note that

$$\epsilon^{j(\delta_k - \delta_k)} = 1.0$$

when $m = k$.

$$P_k + jQ_k = |E_k|^2 G_{kk} - j|E_k|^2 B_{kk}$$

$$+ \sum_{m \in \alpha_k} E_k E_m \epsilon^{j(\delta_k - \delta_m)} (G_{km} - jB_{km}) \tag{7.79}$$

Substitute ψ_{km}, the angle across the line km for the difference of the two angles of the bus voltages.

$$P_k + jQ_k = |E_k|^2 G_{kk} - j|E_k|^2 B_{kk} + \sum_{m \in \alpha_k} E_k E_m \epsilon^{j\psi_{km}} (G_{km} - jB_{km}) \tag{7.80}$$

Remembering that $\epsilon^{j\psi_{km}} = \cos\psi_{km} + j\sin\psi_{km}$, one can write this equation

$$P_k + jQ_k = |E_k|^2 G_{kk} - j|E_k|^2 B_{kk} + \sum_{m \in \alpha_k} E_k E_m (\cos\psi_{km} + j\sin\psi_{km})(G_{km} - jB_{km}) \tag{7.81}$$

Performing the multiplication and collecting the real and imaginary parts, equation 7.82 is obtained.

$$P_k + jQ_k = |E_k|^2 G_{kk} - j|E_k|^2 B_{kk} + \sum_{m \in \alpha_k} E_k E_m [(G_{km} \cos\psi_{km} + B_{km} \sin\psi_{km})$$

$$+ j(G_{km} \sin\psi_{km} - B_{km} \cos\psi_{km})] \tag{7.82}$$

Separation of real and imaginary parts gives equations 7.83 and 7.84.

$$P_k = |E_k|^2 G_{kk} + \sum_{m \in \alpha_k} E_k E_m (G_{km} \cos \psi_{km}$$

$$+ B_{km} \sin \psi_{km}) \tag{7.83}$$

$$Q_k = -|E_k|^2 B_{kk} + \sum_{m \in \alpha_k} E_k E_m (G_{km} \sin \psi_{km}$$

$$- B_{km} \cos \psi_{km}) \tag{7.84}$$

DEVELOPMENT OF THE REAL POWER MODEL

In equation 7.83 substitute $1 + \cos \psi_{km} - 1$ for $\cos \psi_{km}$ and $\psi_{km} + \sin \psi_{km} - \psi_{km}$ for $\sin \psi_{km}$:

$$P_k = |E_k|^2 G_{kk} + E_k \sum_{m \in \alpha_k} E_m \{ G_{km} (1 + \cos \psi_{km} - 1)$$

$$+ B_{km} (\psi_{km} + \sin \psi_{km} - \psi_{km}) \} \tag{7.85}$$

which can be rewritten as the sum of two summations, by splitting the last term on the right into two terms, $B_{km} \psi_{km} + B_{km} (\sin \psi_{km} - \psi_{km})$.

$$P_k = |E_k|^2 G_{kk} + E_k \sum_{m \in \alpha_k} E_m \{ G_{km} (1 + \cos \psi_{km} - 1)$$

$$+ B_{km} (\sin \psi_{km} - \psi) \} + E_k \sum_{m \in \alpha_k} E_m B_{km} \psi_{km} \tag{7.86}$$

This equation can be rearranged and solved for

$$E_k \Sigma E_m B_{km} \psi_{km}.$$

$$E_k \Sigma E_m B_{km} \psi_{km} = P_k - |E_k|^2 G_{kk}$$

$$- E_k \sum_{m \in \alpha_k} E_m G_{km} - E_k \sum_{m \in \alpha_k} E_m \{ G_{km} (\cos \psi_{km} - 1) + B_{km} (\sin \psi_{km} - \psi_{km}) \}$$

$$\tag{7.87}$$

It was pointed out on p. 161 that the angular difference of the voltage across a transmission line is of the order of a few degrees; therefore, the Taylor's expansion for the $\sin \psi_{km}$ and $\cos \psi_{km}$ can be terminated after the second term in the expansion of these functions without adversely affecting the results of the calculation within the accuracy required in contingency analysis.

$$
\left.\begin{aligned}
\sin \psi_{km} &\approx \psi_{km} - \frac{\psi_{km}^3}{6} \\[6pt]
\cos \psi_{km} &\approx 1 - \frac{\psi_{km}^2}{2}
\end{aligned}\right\} \tag{7.88}
$$

Then

$$
G_{km}(\cos \psi_{km} - 1) \approx -\frac{G_{km}\psi_{km}^2}{2}
$$

and

$$
B_{km}(\sin \psi_{km} - \psi_{km}) \approx -B_{km}\frac{\psi_{km}^3}{6} \tag{7.89}
$$

By substituting these approximate values into equation 7.87 and retaining the equal sign, since the equations are true within the accuracy required in contingency analysis, one obtains equation 7.90:

$$
E_k \sum_{m \in \alpha_k} E_m B_{km} \psi_{km} = \left(P_k - |E_k|^2 G_{kk} - E_k \sum_{m \in \alpha_k} E_m G_{km} \right)
$$

$$
+ \left[E_k \sum_{m \in \alpha_k} E_m \left(G_{km}\frac{\psi_{km}^2}{2} + B_{km}\frac{\psi_{km}^3}{6} \right) \right] \tag{7.90}
$$

The opposite substitution that was made in changing equation 7.79 to equation 7.80 is used to convert equation 7.90 to 7.91. In addition, set

$$
P_k' = P_k - |E_k|^2 G_{kk} - E_k \sum_{m \in \alpha_k} E_m G_{km} \tag{7.90a}
$$

and

$$
P_k'' = E_k \sum_{m \in \alpha_k} E_m \left(G_{km}\frac{\psi_{km}^2}{2} + B_{km}\frac{\psi_{km}^3}{6} \right) \tag{7.90b}
$$

then

$$
E_k \sum_{m \in \alpha_k} E_m B_{km}(\delta_k - \delta_m) = P_k' + P_k'' \tag{7.91}
$$

Defining the elements of an N-by-N-matrix by means of the relationships $A_{kk} = E_k \sum_{m \in \alpha_k} E_m B_{km}$ and $A_{km} = -E_k E_m B_{km}$, one can write the complete set of equations for a system as

$$[A][\delta] = [P' + P''] \tag{7.92}$$

which expresses the fact that the real power injection into a bus is a function of the angles of the bus voltages.

DEVELOPMENT OF THE REACTIVE POWER MODEL

In developing the reactive power model, account must be taken of the off-nominal tap transformers. The π equivalent that was described by Ward and Hale [13] for use in nodal analysis of networks is shown in Fig. 7.22, in which a turns ratio device is in series with an admittance y_{km}.

The equivalent network can be represented in terms of the transfer admittance Y_{km}, which is equal to the negative of the effective transformer admittance y'_{km}. The effective transformer admittance y'_{km} is equal to the transformer admittance times the tap ratio,

$$y'_{km} = +t_{km}y_{km} = -Y_{km} \tag{7.93}$$

The ground leg of the equivalent that is connected to bus k can be obtained in terms of the transfer admittance [16].

$$(t_{km} - 1)t_{km}y_{km} = (t_{km} - 1)y'_{km} = -(t_{km} - 1)Y_{km} \tag{7.94}$$

The ground leg of the equivalent that is connected to bus m can be obtained in terms of the transfer admittance.

$$(1 - t_{km})y_{km} = \frac{(1 - t_{km})y'}{t_{km}} = -\frac{1 - t_{km}}{t_{km}}Y_{km} \tag{7.95}$$

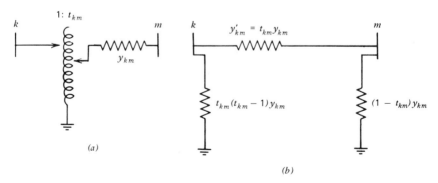

Fig. 7.22. (*a*) Tap transformer. (*b*) Network equivalent.

The contribution of a transformer to the driving point admittance of bus k is obtained by adding equations 7.93 and 7.94. The contribution to the driving point admittance of bus m is obtained by adding equation 7.93 and 7.95 (see Fig. 7.23).

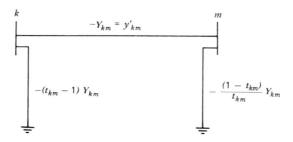

Fig. 7.23.

$$Y_{kk} = - Y_{km} - (t_{km} - 1) Y_{km} = - t_{km} Y_{km} \qquad (7.96)$$

$$Y_{mm} = - Y_{km} - \frac{1 - t_{km}}{t_{km}} Y_{km} = - \frac{Y_{km}}{t_{km}} \qquad (7.97)$$

The term B_{kk} in equation 7.84 can be separated into its component parts:

1. The sum of series susceptances of lines and transformers connected to adjacent buses.

2. The sum of shunt susceptances of ground tie legs of the π equivalents of transformers.

3. The total line charging and shunt capacitors (or reactors) connected to bus k.

$$B_{kk} = \sum_{m \in \alpha_k} (- t_{km} B_{km} + B_{\pi km}) + B_{ck} \qquad (7.98)$$

where B_{km} = the transfer susceptance of the branch km

$\quad\quad t_{km}$ = the tap ratio if the branch is a transformer otherwise $t_{km} = 1$

$\quad\quad B_{\pi km}$ = the charging susceptance of the k leg of the transformer km

$\quad\quad B_{ck}$ = the total susceptance of shunt capacitors connected to bus k

(shunt inductive reactors are negative). Substitution of equation 7.98 and $\cos \psi = 1 + \cos \psi - 1$ into equation 7.84 gives

$$Q_k = E_k^2 \sum_{m \in \alpha_k} (t_{km} B_{km} - B_{\pi km}) - E_k^2 B_{ck} - E_k \sum_{m \in \alpha_k} E_m B_{km}$$

$$+ E_k \sum_{m \in \alpha_k} E_m \{ G_{km} \sin \psi_{km} - B_{km} (\cos \psi_{km} - 1) \} \qquad (7.99)$$

Dividing through the equation by E_k and rearranging gives

$$\sum_{m \in \alpha_k} (E_k t_{km} B_{km} - E_m B_{km}) = \frac{Q_k}{E_k} + E_k \left(\sum_{m \in \alpha_k} B_{\pi km} + B_{ck} \right)$$

$$- \sum_{m \in \alpha_k} E_m \{ G_{km} \sin \psi_{km} - B_{km} (\cos \psi_{km} - 1) \} \quad (7.100)$$

This equation only applies to buses that have a constraint on vars to be supplied to the network, that is, load buses for which P and Q are specified. For generator buses that have P and V prescribed, Q is unconstrained; therefore equation 7.100 does not apply. The number of equations should be reduced to include only load buses. In this reduced set of equations the terms on the left-hand side that apply to generator (PV) buses are transferred to the right-side of the equation. Let α_k be divided into two subsets:

1. The set of branches between k and type PQ buses, η_k.
2. The set of branches between k and type PV buses, μ_k.

Equation 7.100 can then be written

$$E_k \sum_{m \in \alpha_k} t_{km} B_{km} - \sum_{m \in \eta_k} E_m B_{km}$$

$$= \frac{Q_k}{E_k} + E_k \left(\sum_{m \in \alpha_k} B_{\pi km} + B_{ck} \right) - \sum_{m \in \alpha_k} E_m \{ G_{km} \sin \psi_{km}$$

$$- B_{km} (\cos \psi_{km} - 1) \} + \sum_{m \in \mu_k} E_m B_{km} \quad (7.101)$$

Define

$$Q'_k = \frac{Q_k}{E_k} + E_k \left(\sum_{m \in \alpha_k} t_{km} B_{\pi km} + B_{ck} \right) + \sum_{m \in \mu_k} E_m B_{km} \quad (7.102)$$

and

$$Q''_k = - \sum_{m \in \alpha_k} E_m \left\{ G_{km} \left(\psi_{km} - \frac{\psi_{km}^3}{6} \right) + B_{km} \frac{\psi_{km}^2}{2} \right\} \quad (7.103)$$

Note: Equations 7.88 have been used to replace $\sin \psi$ and $\cos \psi$. Equation 7.101 can then be

$$E_k \sum_{m \in \alpha_k} t_{km} B_{km} - \sum_{m \in \eta_k} E_m B_{km} = Q'_k + Q''_k \quad (7.104)$$

The complete set of equations for the system can then be written in sparce matrix form by defining $C_{kk} = \sum_{m \in \alpha_k} t_{km} B_{km}$ and $C_{km} = -B_{km}$ (for $m \in \eta_k$)

$$[C][E] = [[Q'] + [Q'']]$$ (7.105)

A power flow problem using the decoupled Newton-Raphson method requires the solution of the two sets of equations 7.92 and 7.105 which are repeated here.

$$[A][\delta] = [P' + P'']$$ (7.92)

$$[C][V] = [Q' + Q'']$$ (7.105)

The sparcity of the matrices $[A]$ and $[C]$ must be exploited to achieve speed of solution and economy of memory. The sparcity of the matrix $[A]$ is identical to that of the nodal admittance matrix of the base case power flow while the matrix $[C]$ is more sparce than $[A]$, since it corresponds to the rows and columns of only the fixed P and fixed Q buses of the nodal admittance matrix. Several schemes are available for triangularization of these two matrices:

1. Use the same ordering routine for both matrices and sacrifice speed for simplicity of programming.

2. Matrix $[C]$ is more sparce than $[A]$ and advantage of this fact could be exploited to achieve a higher degree of conservation of sparcity.

3. Order the matrix $[A]$, but at any point in the ordering where there is an equal choice in selecting the next bus in the ordering routine, take the matrix $[C]$ into account in making the decision.

In choosing an ordering scheme, one must balance gain in speed in the solution and the extra burden of a more sophisticated ordering technique. In practice [7], the same ordering for both matrices has been found to be satisfactory.

Beginning with a base case power flow that has been obtained by the Newton-Raphson method, one must obtain the solution for the network for a contingency condition in which a line has been removed or there has been a change in the generation schedule. Treating the line outage case first, one must modify both the $[A]$ and $[C]$ matrices to reflect the change in the network configuration. To change the schedule of generation one modifies the P and Q vectors of equations 7.92 and 7.105.

To remove a line from the nodal admittance matrix, four changes must be made in elements of the matrix. The admittance of the line being removed must be subtracted from the diagonal elements of the admittance matrix corresponding to the buses at the ends of the line being removed, and added to the two off diagonal elements involved. Thus to remove a line between buses k and m whose admittance is y_{km}, the modifications to be made to the four elements of the admittance matrix are:

$$
\left.
\begin{aligned}
Y'_{kk} &= Y_{kk} - y_{km} \\
Y'_{mm} &= Y_{mm} - y_{km} \\
Y'_{km} &= Y_{km} + y_{km} \\
Y'_{mk} &= Y_{mk} + y_{km}
\end{aligned}
\right\}
\qquad (7.106)
$$

Here Y_{kk} indicates an element of the original nodal admittance matrix; Y'_{kk} is the modified element and y_{km} is the admittance of the line being removed.

In matrix notation these modifications of the matrix A can be expressed as

$$
[A'] = [A] - y_{km}[M_A][M_A]^T
\qquad (7.107)
$$

If $[A]$ is an N-by-N-matrix, then $[M_A]$ is an N element vector in which all elements are zero except for the kth and mth elements, and they are $+1$ and -1 respectively. For the outage of a line, the equations that must be satisfied are obtained by substitution of equation 7.107 into equation 7.92, and the angles must be augmented by $\Delta\delta$ to account for the change in the angles of the voltages of the system.

$$
\left[[A] - y_{km}[M_A][M_A]^T\right][\delta + \Delta\delta] = [[P'] + [P'']]
\qquad (7.108)
$$

The angles are the unknowns that must be determined and can be obtained by premultiplying both sides of the equations 7.108 by the inverse of the modified matrix of equation 7.107.

$$
[\delta + \Delta\delta] = \left[[A] - y_{km}[M_A][M_A]^T\right]^{-1}[[P'] + [P'']]
\qquad (7.109)
$$

To achieve maximum speed, one must triangularize matrix $[A]$ (and also $[C]$) only once in a series of contingency studies. It is therefore necessary to compute the modifications that must be made to the triangularized matrix $[A]$ for the removal of a line without resorting to retriangularization [17]. This is achieved by obtaining the inverse of the matrix $[[A] - y_{km}[M_A][M_A]^T]$. In a paper by Peterson, Tinney, and Bree [7] this inverse is given as a series of matrix operations.

Since the author has used the impedance matrix building algorithm extensively in previous chapters, the modification of the inverse for the removal of a line is developed from that point of view and the relationship to the work of reference [7] pointed out at each step in the development. The changes that must be made to the matrix $[A]^{-1}$ when an impedance is added between buses k and m must be found. (*Note*: Removal of a line is achieved by adding a line whose impedance is the negative of the original line incorporated in the system. Thus a line removal is achieved by the line addition routine.) In the building algorithm to add a loop closing line, the matrix is augmented by a column equal to the difference of the elements of column k and column m; $Z_{n\text{-loop}} = (Z_{nk} - Z_{nm})$, but this difference can be expressed as:

$$[A]^{-1}[M_A] = [Z_{kn} - Z_{mn}] = [Z_{n\text{-loop}}] \tag{7.110}$$

The matrix $[A]^{-1}$ is augmented by a row equal to the difference of the elements of rows k and m (see equation 3.13). This row vector is expressed as:

$$[M_A]^T[A]^{-1} = [Z_{nk} - Z_{nm}] = [Z_{\text{loop}-n}] \tag{7.111}$$

The diagonal element of the new axis being added to the matrix $[A]^{-1}$ is obtained by means of equation 3.12.

$$Z_{\text{loop-loop}} = z_{km} + Z_{kk} + Z_{mm} - 2Z_{km}$$

in which Z_{kk}, Z_{mm}, and Z_{km} are elements from the matrix, and z_{km} is the impedance of the line being added. It can be verified that the diagonal element can also be expressed in terms of the matrices A^{-1} and M_A as

$$Z_{\text{loop-loop}} = z_{km} + [M_A]^T[A]^{-1}[M_A] \tag{7.112}$$

The augmented matrix is therefore as shown in Fig. 7.24.

$$\begin{array}{|c|c|}
\hline
A^{-1} = Z_1 & A^{-1}M_A = Z_2 \\
\hline
M_A^T A^{-1} = Z_3 & z_{km} + M_A^T A^{-1} M_A = Z_4 \\
\hline
\end{array}$$

Fig. 7.24.   Relationship is shown between Peterson and Kron.

The inverse of the matrix that represents the system after adding the line whose impedance is z_{km} is obtained by eliminating this extra axis by means of a Kron reduction (see equation 3.15). Substitution of the values for Z_1,

Z_2, Z_3, and Z_4 from Fig. 7.24 into equation 3.15 gives equation 7.113.

$$[A']^{-1} = [A]^{-1} - [A]^{-1}[M_A]\left[z_{km} + [M_A]^T[A]^{-1}[M_A]\right]^{-1}[M_A]^T[A]^{-1}$$

$$(7.113)$$

Since $z_{km} + [M]^T[A]^{-1}[M_A]$ is a single element, it is a scalar and the order in which it is included in the multiplication need not be maintained. The equation can therefore be written:

$$[A']^{-1} = [A]^{-1} - \left[\left[z_{km} + [M_A]^T[A]^{-1}[M_A]\right]^{-1}\right][A]^{-1}[M_A][M_A]^T[A]^{-1}$$

$$(7.114)$$

Equation 7.114 is taken from Ref. 7 for the change to be made in the inverse matrix for the removal of a line from the system. We have seen that it is the Kron reduction $Z_1' = Z_1 - Z_2 Z_4^{-1} Z_3$. Substitution of equation 7.114 into equation 7.109 gives

$$(\delta + \Delta\delta) = \left[[A]^{-1} - \left\{z_{km} + [M_A]^T[A]^{-1}[M_A]\right\}^{-1}[A]^{-1}[M_A][M_A]^T[A]^{-1}\right]$$

$$\times[[P' + P'']] \qquad (7.115)$$

Multiplication is distributive; therefore, this equation may be expanded to read:

$$[\delta + \Delta\delta] = \left[[A]^{-1}[[P'] + [P'']]\right]$$

$$- \left\{z_{km} + [M_A]^T[A]^{-1}[M_A]\right\}^{-1}\left[[A]^{-1}[M_A][M_A]^T[A]^{-1}\right][[P'] + [P'']]$$

$$(7.116)$$

From

$$[A][\delta] = [[P'] + [P'']]$$

as given in equation 7.92,

$$[\delta] = [A]^{-1}[[P'] + [P'']] \qquad (7.117)$$

Substitution of equation 7.117 in two places in equation 7.116 gives

$$[\delta+\Delta\delta]=[\delta]-\left\{z_{km}+[M_A]^T[A]^{-1}[M_A]\right\}^{-1}[A]^{-1}[M_A][M_A]^T[\delta]$$

$$(7.118)$$

Substracting $[\delta]$ from both sides of the equation gives

$$[\Delta\delta]=-\left\{z_{km}+[M_A]^T[A]^{-1}[M_A]\right\}^{-1}[A]^{-1}[M_A][M_A]^T[\delta] \quad (7.119)$$

Since $[M_A]$ is a vector, all of its elements are zero except the kth and mth elements, which are equal to $+1$ and -1, respectively. Then

$$[M_A]^T[\delta]=(\delta_k-\delta_m) \tag{7.120}$$

For simplification of the notation define

$$[Z_A]=[A]^{-1}[M_A] \tag{7.121}$$

Premultiplication of this equation by $[M_A]^T$ will result in the difference of two scalar elements from $[Z_A]$, as shown in equation 7.122.

$$[M_A]^T[Z_A]=[M_A]^T[A]^{-1}[M_A]\equiv(Z_{A_k}-Z_{A_m}) \tag{7.122}$$

Incorporating both equations 7.121 and 7.122 into equation 7.119 will result in equation 7.123.

$$[\Delta\delta]=-\left(z_{km}+Z_{A_k}-Z_{A_m}\right)^{-1}(\delta_k-\delta_m)[Z_A] \tag{7.123}$$

Although $[Z_A]$ is an explicit inverse of $[A]$, it should be obtained from the triangularized form of $[A]$.

The same series of steps can be used to obtain the voltage magnitude adjustments produced by the line outage. The steps will be given without the detailed explanation that was given for the determination of the change of the voltage angular position. The matrix $[C]$ is modified for the line outage

$$[C']=[C]-y_{km}[M_C][M_C]^T \tag{7.124}$$

$$\left[[C]-y_{km}[M_C][M_C]^T\right][E+\Delta E]=[[Q']+[Q'']] \tag{7.125}$$

$$[E+\Delta E]=\left[[C]-y_{km}[M_C][M_C]^T\right]^{-1}[[Q']+[Q'']] \qquad (7.126)$$

$$[E+\Delta E]=\left[[C]^{-1}-\left\{z_{km}+[M_C]^T[C]^{-1}[M_C]\right\}^{-1}[C]^{-1}[M_C][M_C]^T[C]^{-1}\right]$$
$$\times[[Q']-[Q'']] \qquad (7.127)$$

$$[E+\Delta E]=\left[[C]^{-1}[[Q']+[Q'']]\right]-\left\{z_{km}+[M_C]^T[C]^{-1}[M_C]\right\}^{-1}$$
$$\times\left[[C]^{-1}[M_C][M_C]^T[C]^{-1}\right][[Q']+[Q'']] \qquad (7.128)$$

$$[E+\Delta E]=[E]-\left\{z_{km}+[M_C]^T[C]^{-1}[M_C]\right\}^{-1}[C]^{-1}[M_C][M_C]^T[E]$$
$$\qquad (7.129)$$

Here M_C will be defined the same as M_A if the outage is a transmission line, but if the outage is a transformer, the kth element is $\sqrt{t_{km}}$ and the nth element is $-1/\sqrt{t_{km}}$ to account for the π legs of the transformer. If the line to be removed is between a PQ bus and a voltage regulated bus, a single entry of 1 is required for the kth element of the vector $[M_C]$. If the line outage is between two voltage regulated buses, equation 7.129 does not apply. Adding $[E]$ to both sides of equation 7.129

$$[\Delta E]=-\left\{z_{km}+[M_C]^T[C]^{-1}[M_C]\right\}^{-1}[C]^{-1}[M_C][M_C]^T[E] \qquad (7.130)$$

METHOD OF APPLICATION

The main steps that are required in the analysis of a set of contingencies of a base system are as follows:

1. Describe the transmission network; supply the initial estimate of the voltage profile (or a converged voltage solution), and give the load and generation of the base case. Describe the series of contingencies to be investigated. If a generator is to be removed, supply the distribution of the generation to the remaining generators.

2. Form the matrices $[A]$ and $[C]$ in triangularized form using equations 7.92 and 7.105, respectively.

3. Compute $[P']$ using equation 7.90a and the voltage magnitudes from step 1.

4. Modify P_k' and P_m' for the effect of the line removal using equation 7.90a.

5. By means of equation 7.121 and the matrix $[M]$ for the particular outage, solve for the vector $[Z_A]$ by ordered elimination.

6. Compute the angular corrections $[\Delta\delta]$ using formula 7.119 and determine the new angular position of the bus voltages,

$$[\delta'] = [\delta + \Delta\delta] \qquad (7.131)$$

and compute P'' for the new angles using equation 7.90b.

The revised voltage magnitudes are computed by a similar set of calculations.

COMPARISON OF THE Z-MATRIX AND NEWTON-RAPHSON METHOD

The Newton-Raphson method has certain advantages over the Z-matrix method. Every line of the system is checked for overload conditions and every bus can be tested for low voltage. Thus the system planner or the system operator can be completely ignorant of his system and the program will find the lines that are in trouble. This feature is especially helpful for future studies when planners may have lost their "feel" for their system. The Newton-Raphson method is somewhat more accurate than the Z-matrix method, since the contingency is iterated to a prescribed tolerance. Perhaps the accuracy is too high for the purpose when the uncertainties of the future system is taken into account. The decoupled Newton-Raphson method is reported to be six times faster than the full Newton-Raphson method. What the Z-matrix lacks in accuracy, it makes up in speed, since contingencies (after the first case) are evaluated at 500 times load flow solution speed.

DISTRIBUTION FACTORS

Another method of evaluating contingencies that must be mentioned is the method of distribution factors. A distribution factor is the fraction of the flow in a line that is transferred to another line when the first line is opened. This method has the advantage that a contingency from a base case can be computed for the actual flow on a line in a real life situation. It has the disadvantage that many distribution factors that are computed and saved in a tabulation (computer output or memory) will never be used. Furthermore, beginning with a base case system, a distribution factor for the first contingency must modify the distribution factor of the second contingency (reminding the author of the saying: "Little flees, have lesser flees upon their backs to bite them, and these in turn have lesser still and so on ad infinitum"). The method must not be taken lightly however because it does have definite applications in practice (see the discussions in Ref. 5 by C. A. Falcone and by Walter L. Synder Jr.).

References

1. **C. A. MacArthur,** Transmission limitations computed by superposition, *Trans. AIEE PA &S.,* Vol. 57, (December 1961), p. 827.

2. **A. H. El-Abiad and G. W. Stagg,** Automatic evaluation of power system performance—Effects on line and transformer outages, *Trans. AIEE PA & S,* Vol 64, (February 1963), p. 712.

3. **H. C. Limmer,** Techniques and applications of security calculations applied to dispatching computers, *PSCC Proc.,* Rome, Italy, 1969.

4. **H. E. Brown,** Contingencies evaluated by a Z-matrix method, *IEEE Trans. PA & S,* Vol. 88, (April 1969), p. 409.

5. **H. E. Brown,** Interchange capability and contingency evaluation by a Z-matrix method, *IEEE Trans. PA & S,* Vol. 91, (September-October 1972), p. 1827.

6. **G. L. Landgren, H. L. Terhune, and R. K. Angel,** Transmission interchange capability—Analysis by computer, *IEEE Trans. PA & S,* Vol. 91, (November-December 1972), p. 2405.

7. **N. M. Peterson, W. F. Tinney, and D. W. Bree, Jr.,** Iterative linear a.c. power flow solution for fast approximate outage studies, *IEEE Trans. PA & S,* Vol 91, (September-October 1972), p. 2048.

8. **S. T. Despotovic, B. S. Babic, and V. P. Mastilovic,** A rapid and reliable method for solving load flow problems, *IEEE Trans. PA & S,* Vol. 90, (January-February 1971), p. 123.

9. **B. Stott,** Decoupled Newton load flow, *IEEE Trans. PA & S,* Vol. 91, (September-October 1972), p. 1955.

10. **H. E. Brown and C. E. Person,** Short circuit studies of large systems by the impedance matrix method, *IEEE PICA Conf. Proc.,* Pittsburgh, 1967, p. 335.

11. **H. E. Brown, L. K. Kirchmayer, C. E. Person, and G. W. Stagg,** Digital calculation of 3-phase short-circuits by matrix method, *Trans. AIEE,* Vol 79, Part 3, (February 1961), p. 1277.

12. **H. E. Brown, G. K. Carter, H. H. Happ, and C. E. Person,** Power flow solution by impedance matrix iterative method, *Trans. AIEE,* Vol 82, Part 3, (1963), p. 1.

13. **J. B. Ward and H. W. Hale,** Digital computer solution of power-flow problems, *Trans. AIEE,* Vol 75, Part 3, (June 1956), p. 398.

14. **H. E. Brown,** Simultaneous interchange optimization by means of the Z-matrix, *IEEE PICA Conf. Proc.,* 1973.

15. **G. T. Heydt and S. W. Anderson,** Distribution factors from Z bus matrix for interchange capability analysis, *IEEE Winter Meeting,* 1973.

16. **J. P. Britton**, Improved load flow performance through a more generalized equation form, *IEEE Trans. PA & S*, Vol 90, (January 1971), p. 109.

17. **A. Ralston and H. Wilf**, *Mathematical Methods for Digital Computer*, Wiley, 1960, pp. 73–77.

18. **A. F. Glimn and G. W. Stagg**, Automatic calculation of load flows, *Trans. AIEE*, Vol 76, Part 3, (1957), p. 817.

19. **J. E. Van Ness and J. H. Griffin**, Elimination methods for load flow studies, *Trans. AIEE PA & S*, Vol. 80, (June 1961), p. 299.

20. **N. Sato and W. F. Tinney**, Techniques for exploiting the sparcity of the network admittance matrix, *Trans IEEE PA&S*, Vol 82, (December 1963), p. 944.

21. **W. F. Tinney and C. E. Hart**, Power flow solutions by Newton's method, *Trans. IEEE PA&S*, Vol. 86, (November 1967), p. 1449.

22. **J. Carpentier**, Application of Newton's method to load flow problems, *Proc. Power Systems Computer Conf.*, London, 1963.

8

Transient Stability

Transient stability involves the dynamic behavior of a power network that has been in a balanced steady-state power flow condition and is subjected to a disturbance by either a momentary short circuit, the loss of a major transmission line, or the loss of a generator or a sudden increase in load. The disturbance unbalances the network because a number of the variables or parameters in the power flow equations that determine the power supplied to the network by the generators are changed by the disturbance. The loss of a transmission line modifies the effective admittance between the several generators Y_{km}; a short circuit modifies the voltage magnitudes and angles of the buses, and the loss of a generator produces an unbalance between generation and load connected to the system.

Since the mechanical (steam or hydro) input to a turbine can not be altered instantaneously to match the change in electrical output of its generator, the difference between input and output power, produced by the disturbance, results in an accelerating torque being applied to the rotor. Because the magnitudes of the inertia of the generators are unequal as is the amount of power available for acceleration of the several machines, they change their angular velocity and position, with reference to a synchronously rotating vector, at different rates as required to satisfy the set of differential equations 8.1.

$$M_k \frac{d^2 \delta_k}{dt^2} = \left(P_{M_k} - P_{E_k} \right) = P_{A_k} \tag{8.1}$$

where P_{M_k}, P_{E_k}, and P_{A_k} are the mechanical input power, the electrical output power, and the accelerating power of generator k. The M_k is the inertia of the rotor of the generator and prime mover, and δ_k is the angular displacement of the generator's internal voltage with respect to a reference vector rotating at synchronous speed. For each generator represented in

178

the system there is an equation in the set 8.1.

Any change in the relative angular displacement of the generator rotors (voltages) with respect to each other will produce a variation in the electrical power they individually deliver to the system. The change in power will, in turn, alter the accelerating power used in equation 8.1. Therefore, the problem involves solving the set of differential equations for the new angular position of the generator voltages after the accelerating power has acted on the rotors for a small increment of time ΔT and these new angles have been substituted in the load flow algebraic equations to determine new power output values for use in the differential equations during the next time interval. The swing curve solution for the simplest machine representation therefore involves alternately solving the network algebraic equations and the differential equations of motion for the duration of the study at time steps of 0.05 to 0.10 sec. A single study simulates up to 2 sec. of real time (more detailed representations are used to simulate up to 10 sec.).

The second-order differential equations of motion 8.1 are solved by performing two integrations [1],

$$\frac{d^2\delta}{dt^2} = \frac{d\omega}{dt} = \frac{Pa}{M} \tag{8.2}$$

In which ω is the angular velocity of the rotor with respect to the synchronous reference vector. The first integration gives equation 8.3.

$$\frac{d\delta}{dt} = \omega = \omega_0 + \frac{P_a t}{M} \tag{8.3}$$

$$\delta = \omega t = \delta_0 + \omega_0 t + \frac{P_a t^2}{M} \tag{8.4}$$

A more elaborate generator model that includes amortisseur windings in the direct and quadrature axes will require from 200 to 500 solutions/psec of real time because of smaller time constants in some of the circuits. The more detailed model needs as much as 10 more items of data for each generator represented in this degree of detail and the solution of as many as 20 differential equations per generator for each time step in the simulation of the voltage regulator, exciter, and flux decay in the magnetic circuits.

If the study indicates that after a disturbance a new steady-state condition will be reached, the system is said to be transient stable for the condition being tested. However, increasing the length of time required to clear the fault from the system (stuck breaker with the fault being cleared by back-up protection), applying the fault at a different location, removing a different line, losing a different generator, different timing of reclosing after the fault has cleared, or changing the amount of details being represented in the model may cause the system to be unstable. Therefore, a great many studies must be made to ensure that the system is secure (stable) for the contingencies that could reasonably be expected to occur.

Since entire books have been written on transient stability [1–3], numerical methods of solution of differential equations [4,5], and control systems [6], the present work only discusses some aspects of the problem as they relate to digital computation.

DEVELOPMENT OF THE TECHNIQUES

After the nodal iterative method of computing power flows was described in 1956, the load flow program size increased quickly from six buses to 200 buses. During the several years that the power flow programs were stabilized at this size, every generator of a system could be represented in transient stability studies. The digital computer representation of the machines evolved during this period and soon surpassed the sophistication used in the modeling of other parts of the system. Customer loads were still represented by constant impedances connected to ground when the programs could include transient saliency, voltage regulators, governors, damper windings, and such [7,8].

Even after methods had been developed to represent loads as constant Z, constant MVA, and constant I, little was actually known about the true nature of the system loads, and, because of this uncertainty, these complicated load representations were seldom used. Some work was done [9,10] by making tests on isolated feeders to obtain the relationship between load and voltage, but the characteristics obtained could not be universally applied to other feeders. There is some doubt if the feeders that had been tested would yield the same results if they were retested at a later time.

With the increase in EHV interconnections the size of the studies increased steadily to 1000, 2500, and 4500 bus systems. Preparation of data for the detailed model of the generators for the number of machines represented in a 2500-bus system (in the order of 700 generators) is very difficult and is certainly unwarrented. Only those generators that are electrically closely coupled to the disturbance should be represented in this detail.

In his discussion of Ref. 11, Kimbark points out that all of the machine representation refinements increased accuracy from 1 to 2% (for the system being studied) over the so-called classical method in which only constant voltage behind transient reactance is assumed. He points out that the voltage regulator representation [13] requires five constants that are hard (if not impossible) to obtain, and suggests that such refinements should only be used in very critical cases [12].

THE DILEMMA

It has been indicated that a transient stability study is very time consuming because a load flow and sets of from 2 to 20 differential equations must be solved, between 100 and 5000 times for each case, depending on the amount of detail included in machine representation and the length of time being simulated. Because of the transmission network changes required for each new generator added, several (many) transient stability cases should be tested. Every additional detail in the representation of the machines or the network should increase the accuracy, but at the expense of increased computer running time and the burden of additional data. In a tutorial paper Tinney [13] points out that the gain in accuracy may be open to question because of the possibility that errors in the parameters of the model could easily offset any gains in accuracy in a more detailed generator model. In the absence of real control parameters, a practice has developed to estimate the values to be used. Furthermore, the engineer making the study may be unaware that control settings in the field may have been altered by a well meaning technician to give a "better" control action; thus the study may not be realistic.

In a companion paper in the tutorial course, Young [13] points out that the representation must be as simple as possible, because the representation may be unnecessarily complex. He suggests that as many simplifying assumptions as possible should be made, especially if some of the data must be estimated as indicated earlier.

deMello [13] indicates that it is important to use skill in making these simplifications. Consideration must be given not only to simplification of the generator model but also to the reduction of the number of generators represented. Generators that are remote from the disturbance can often be replaced by a fixed impedance to ground without affecting the accuracy of the study.

The amount of details included in a transient stability study must be balanced. An increase in the details represented greatly increases the burden of data preparation, the computer memory requirement, and the computer running time per case, in the hope of a small increase of accuracy that may result. Since the computer running time must be limited, increasing the details results in fewer cases tested. As many conditions as possible should be studied, because the case that is not tried for lack of time could be the case that revealed system instability; however, unwise sacrifice of details may result in a study unrealistic and worthless.

The remainder of this chapter outlines the several machine models that can be used, the equations that must be solved, the data requirements for the different models, and the limitations imposed by these machine representations. The discussions of the models are brief.

Since the load flow solutions require at least half of the computer running time, this aspect of the problem is also discussed, including the network equivalent method [1, 16–18] that achieves greater speed in this part of the problem. A brief discussion of the simplification of the system by grouping generators into groups of "coherent" machines and thus reducing the number of generators in the study will also be included [15]. For more detailed treatment of the subject the reader is referred to the considerable source material [1–3, 7–9].

CLASSICAL REPRESENTATION OF A GENERATOR

In a transient stability study the network is first brought into load flow balance by one of the techniques described in Chapter 4. The terminal conditions of the generator have therefore been determined; that is, the angles α of the generator terminal voltages with respect to a reference axis (swing machine terminal), E_t the magnitude of the terminal voltages, and I_t the current injected into the network by the generators are all known.

It is assumed in the "classical" model that the field flux linkages e'_q are constant in magnitude for the duration of the study. Since transient saliency is neglected in this model, $X_q = X'_q = X'_d$ and $e'_q = e' = E_Q$.

Here X_q = quadrature axis synchronous reactance
X'_q = quadrature axis transient reactance
X'_d = direct axis transient reactance
e'_q = field flux linkages
e' = voltage behind transient reactance
E_Q = voltage behind synchronous reactance

An additional bus is added to the network for each generator being represented, and the generator transient reactance is inserted between the generator terminal and the new bus that is inside the machine. The voltages e'_q of the new buses behind the transient reactances are computed by means of equation 8.5, and the angles δ are determined from the vector relationships of Fig. 8.1.

$$e'_q = E_t + I_t jX'_d \qquad (8.5)$$

Throughout the swing curve the magnitude of e'_q remains constant and δ changes at each time step as required by equation 8.1.

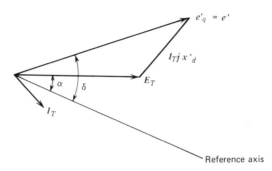

Fig. 8.1. Classical representation of a generator.

The power flow algorithm for use in the swing curve calculation must be modified, because the generator terminal buses no longer are constant voltage buses, it is the buses behind the reactances X_d' that are assumed to have a voltage of constant magnitude. Furthermore, the generators no longer supply constant power P to the system, but at each time step the angle δ of the voltage e_q' is prescribed, and the generator powers injected into the system are the unknowns of primary importance to be determined by the load flow calculation. If all loads are represented as constant impedance loads, the general equation for determining the bus voltages for power flow calculations, as given in equation 8.6, simplifies to equation 8.7, because the power supplied to the load becomes a part of the network flow and is represented by the flow in the element Y_{k-0} rather than by a direct power injection into the system (see Fig. 8.2).

$$E_k = \frac{-\Sigma E_m Y_{km} + (P_k - jQ_k)/E_k^*}{Y_{kk}} \tag{8.6}$$

$$E_k = \frac{-\Sigma E_m Y_{km}}{Y_{kk} + Y_{k-0}} \tag{8.7}$$

Equation 8.6 is quadratic in E_k, since E_k^* appears on the right-hand side of the equation; therefore, an iterative procedure is required to solve for the system bus voltages. Equation 8.7 is linear, and the solution is obtained by a noniterative Gaussian elimination in which sparce matrix methods should be used. In those cases in which loads are represented by constant I and (or) constant MVA, equation 8.6 must be used and the solution is most efficiently obtained by the sparce matrix method of Tinney [19,20].

The classical method is the simplest model that can be used in representing a machine in a transient stability study. Only one additional item of

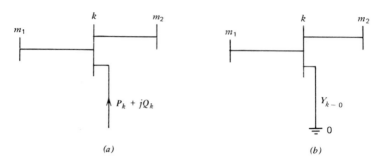

Fig. 8.2. (*a*) Load represented by constant power injection. (*b*) Load represented by constant impedance.

data X_d' is required for each generator. This is the fastest method because the number of load flows that must be calculated per second of real-time simulation is only limited by the rate of change of the angular position of the generators. A maximum change in angular position of 15 degrees in a time step Δt is considered to be satisfactory. Therefore, Δt can usually be between 0.05 and 0.1 sec. If Δt must be reduced below 0.01 sec to keep the generators within the 15 degree limitation, one or more of the generators probably have become unstable and the study should be terminated. In some of the other models to be discussed later, shorter time constants limit Δt to a smaller step size and therefore more load flows must be made per second of real-time simulation.

The classical method is more accurate than might be expected [2] because of compensating effects. When a system is subjected to a short circuit, the terminal voltages of the generators are generally reduced. Thus the voltage differences between the constant voltages behind the transient reactances and the generator terminals are greater, and the vars supplied by the generators therefore increase. The increase in vars tends to decrease e_q' because of increased armature reaction, but fast acting voltage regulators increase the excitation. The two effects are in opposition and e_q' is maintained more or less constant. The method gives good results provided that the short circuit is of short duration (0.1 sec or less) and provided that the short circuit is not too severe. (What constitutes a short circuit that is too severe for the classical method is a matter of judgement—there is no substitute for experience.)

The classical method is generally limited to first or at most second swing studies which involve 2 sec of simulation. If the intention of the study is to determine the effect of negative damping produced by voltage regulators, the method can not be used, since voltage regulators are not explicitly represented. Other studies that require longer swing curves should also use a more sophisticated machine representation.

The majority of the transient stability studies that have been made in the past used the classical machine representation, and presumably it will continue to be used for most studies. The greater speed and the simplicity of the data preparation influence the decision to use this model. As long as it is used within its limitations of short duration faults and for first swing calculations, it is very satisfactory. The greatest defect of the model is the inability to study the effect of excitation and voltage regulators on the stability of the system, since regulators are not represented. Constant e'_q implies the use of regulators that counteract the increase in armature reaction when the system is disturbed, but regulators are not explicitly defined in the model. It is therefore impossible to state whether the model gives optimistic (predicts a higher stability limit than actually exists) or pessimistic results.

TRANSIENT SALIENCY

Transient saliency is relatively unimportant for studies in which the generators are overexcited and therefore supply vars to the system, but not for generators operating at unity or leading power factor because saliency can produce a considerable change in the results obtained in the calculation of the stability margin.

Prior to the application of a disturbance, the terminal conditions of the generators I_t, the current injected into the system, and the terminal voltage E_t have been determined by the base case power flow. The voltage behind the quadrature axis synchronous reactance E_Q is obtained by adding the reactance drop jI_tX_q to the terminal voltage as shown in Fig. 8.3.

The vector E_Q establishes the generator's quadrature axis that is 90 electrical degrees ahead of the direct axis of the field winding. The angle δ that the quadrature axis makes with the arbitrary system reference axis is also determined. The voltage e' behind the direct axis transient reactance X'_d is obtained by adding $jI_tX'_d$ to the terminal voltage as shown in Fig. 8.4.

The projection of e' on the quadrature axis defines e'_q, the voltage that is proportional to the field flux linkages. The e'_q is assumed to be constant if transient saliency is represented. During a disturbance, currents are induced in the damper windings and eddy currents are induced in the iron circuits. These currents oppose change in the various magnetic fluxes. Because of the resistance in the current paths, these currents die out rather quickly, and only the magnetic linkages e'_q remains constant, since it alone is supported by the main field winding. This is the basic assumption on which transient saliency is founded.

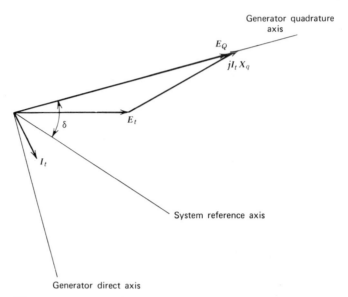

Fig. 8.3.

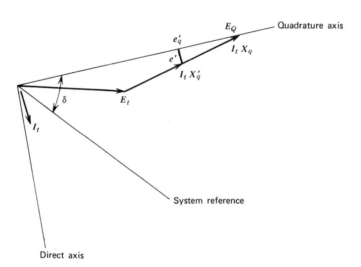

Fig. 8.4. Generator model with transient saliency included.

After a disturbance has been applied to the network a new voltage profile for the system is obtained by the power flow algorithm. In the case of a short circuit, the voltage E_F of the faulted bus is set equal to zero and the voltages of the other buses of the system are determined. The voltages E_Q of the generators are held constant in the load flow solution but must be adjusted later so that e_q' remains unchanged both in magnitude and angular position δ. A new generator current is determined by equation 8.8 using the new generator terminal voltage and the previous value of E_Q.

$$I_t^{new} = \frac{(E_Q - E_t^{new})}{jX_q} \tag{8.8}$$

Using this new current value in equation 8.9, E_Q must be so adjusted so that the new value of $\bar{e}_q'$ is equal to e_q' within an allowable tolerance.

$$|\bar{e}_q'| = |E_Q| - \frac{X_q - X_d'}{|E_Q|}\left\{\operatorname{Im}(E_Q)\operatorname{Re}(I_t^{new}) - \operatorname{Re}(E_Q)\operatorname{Im}(I_t^{new})\right\} \tag{8.9}$$

The Im( ) and Re( ) indicate the imaginary and real part of the quantity in the parentheses.

If $|e_q'|$ does not pass the tolerance test 8.10, an adjusted value of E_Q is determined by means of equation 8.11.

$$\left||\bar{e}_q'| - |e_q'|\right| \leqslant \eta \tag{8.10}$$

$$|E_Q^{new}| = |E_Q|\frac{|\bar{e}_q'|}{|e_q'|} \tag{8.11}$$

The calculations 8.9, 8.10, and 8.11 are repeated as often as is necessary before advancing to the next generator.

For any time step in which it has been necessary to iterate in the determination of E_Q for one or more generators, it is necessary to return to the load flow and solve for a new voltage profile in terms of the revised values E_Q. All calculations 8.8 through 8.11 are repeated as often as necessary. After acceptable values of E_Q have been obtained for all generators, the new generator power outputs are computed by equation 8.12.

$$P_t = \operatorname{Re}(E_Q I_t^*) \tag{8.12}$$

These values of power output are substituted into the differential equations 8.1 and solved for new values of δ to be used in the next time interval Δt.

Figure 8.5 shows the vector diagram before and after the application of the disturbance and the adjustment that is necessary in the magnitude of E_Q to maintain e'_q unchanged.

To include transient saliency an additional item of data is required for each generator, since X'_d and X_q are used in equation 8.9. The calculation included a small iterative loop—equation 8.8 through equation 8.11 plus extra load flow solutions when required. The gain in accuracy is small except when the generators are operating at leading power factor. The solution requires additional time for computation because of the extra load flow solutions and shorter time steps.

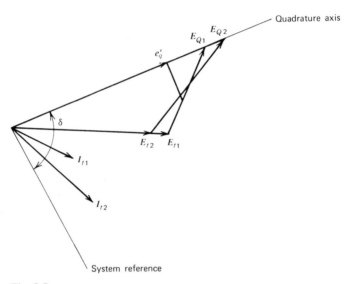

Fig. 8.5.

EXCITATION RESPONSE AND SATURATION

Transient saliency is included in a study to permit the representation of excitation response and saturation effects. The machine model now is extended to include both of these items. After the flux linkages e'_q and the voltage behind the synchronous reactance E_Q have been determined for the prefault condition, it is necessary to find E_I, the steady-state field per unit excitation, or per unit field current as it is referred to by some authors. The vector $I_t X_d$ is added to the terminal voltage of the vector diagram of Fig. 8.5 to obtain the diagram of Fig. 8.6. The vector $I_t(X_d - X_q)$ is

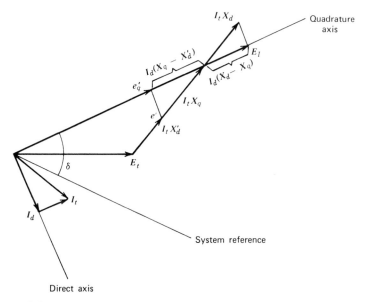

Fig. 8.6.

projected onto the quadrature axis to give $I_d(X_d - X_q)$, where I_d represents the direct axis component of I_t. Equations 8.13 and 8.14 are obtained from the diagram.

$$E_I = E_Q + I_d(X_d - X_q) \tag{8.13}$$

$$e'_q = E_Q - I_d(X_q - X'_d) \tag{8.14}$$

The E_I can be obtained in terms of the known quantities e'_q and E_Q by eliminating I_d between the two equations 8.13 and 8.14.

$$E_I = \frac{(X_d - X'_d)}{(X_q - X'_d)} E_Q - \frac{(X_d - X_q)}{X_q - X'_d} e'_q \tag{8.15}$$

If saturation is not being considered the prefault per unit excitation E_I is equal to the per unit field voltage E_{fd}. When a fault is applied to the system, a new value E_I is obtained by substitution of the direct axis component of the new generator current from equation 8.8 into equation 8.13.

$$E_I^{new} = E_Q + I_d^{new}(X_d - X_q) \tag{8.16}$$

The change in field flux linkages e'_q in the time interval Δt is obtained by solving the differential equation 8.17.

$$\frac{de'_q}{dt} = \frac{(E_{fd} - E_I^{\text{new}})}{T'_{do}} \tag{8.17}$$

The T'_{do} is the direct axis open circuit time constant and E_{fd} is the per unit field voltage.

In the solution of equation 8.9 for determining E_Q, the value e'_q is no longer a constant for the duration of the study but must be modified for each time step to account for the flux decay $\Delta e'_q$, which will in general reduce e'_q. The $|E_Q|$ is obtained by the same procedure as indicated previously except that e'_q in equation 8.10 is obtained from the solution of equation 8.17 for each time step.

To account for saturation, an index of saturation must be introduced to determine the excitation requirement. One index of saturation that is used very commonly is based on the excitation required to produce the voltage behind Potier reactance (see Fig. 8.7).

$$E_I = e'_q + (X_d - X'_d)I_d + \Delta E_I \tag{8.18}$$

The extra excitation required by saturation of the iron is obtained from the open-circuit saturation curve of Fig. 8.8.

The computational burden has increased to between two and three times that required for the classical model. One difficulty is that even if only one generator in the system is represented in this detail, the program running

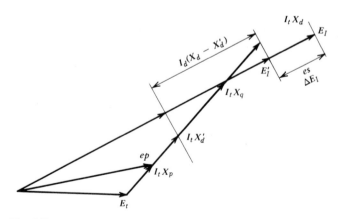

Fig. 8.7.

time is increased two to three times, because the time steps Δt must be reduced. The amount of data to be supplied is increased to include $X_d, X_d', X_q, X_q', T_{do}'$, and an index of saturation. The model can be used to evaluate the effectiveness of voltage regulators and exciter systems. It is also used for swing curves out to 10 sec of real time.

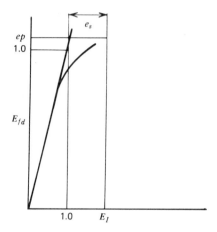

Fig. 8.8.

VOLTAGE REGULATORS

Automatic voltage regulators increase the restoring synchronizing torque in the case of a transient disturbance by forcing greater internal magnetic flux in the generators. It was recognized very early that although voltage regulators could be very helpful in reducing the amplitude of the angles of the first swing, they did contribute negative damping. A point would be reached when further increase in gain and speed of response could cause instability at load levels on the machine that would otherwise be stable if the machine was operating under manual control [21]. Auxiliary signals to the voltage regulators, in addition to the primary signal of generator terminal voltage, have been quite successful in supplying damping to high-speed voltage regulators. The auxiliary signals being used include terminal power, rotor speed, rotor angle, stator current, and rotor current [22].

The complications introduced in the calculations rapidly get out of hand when voltage regulators with auxiliary signals are being represented. Essentially what is being attempted is to vary E_{fd} (see equation 8.17) to increase the restoring torque and increase the stability margin for large angular excursions in the case of a disturbance, and at the same time supply the damping necessary to damp out oscillations in the steady-state mode of operation.

The data requirements include many items, including the regulator characteristics, which are difficult to obtain. A further complication is the fact that the regulator characteristics, rate of response, and ceiling voltage may have been adjusted in the field and unreported to the engineer making the stability study. Some time constants in the various transfer functions [23, 24] are small; thus the allowable time step Δt must be reduced in the calculation. The program speed is greatly reduced, since more time steps per second of simulation are required.

Further refinements in the model to include amortisseur quadrature current, direct axis amortisseur flux, and such cause the program to run so slowly as to be of little value to power systems engineers. These detailed models could produce results of great interest to machine design engineers in the evaluation of damping. These refinements are beyond the sphere of interest of this volume.

LOAD FLOW CALCULATIONS

The calculations can be greatly simplified if all small generators and large generators remote from the disturbance are converted to negative loads. Then all buses except the remaining generators and the fault bus are eliminated from the network by means of the matrix equation 8.19.

$$Y_1' = Y_1 - Y_2 Y_4^{-1} Y_3 \qquad (8.19)$$

In this equation the submatrix Y_1 corresponds to the generator axes and the columns of Y_2 and the rows of Y_3 correspond to the buses to be eliminated [1, 16–18]. See the illustrative example at the end of this section. In the interest of computer speed, the elimination is most advantageously carried out during the assembly of the Y-matrix as indicated in Ref. 18. As each bus is included in the matrix, the elimination of equation 8.19 is performed if the bus is not a generator bus. The advantage is that the inverse of matrix Y_4 becomes merely a division of vector Y_2 by a scalar and the eliminations begin when the matrix is very small. This is equivalent to the technique used in forming the Z-matrix by the building algorithm, line by line, and eliminating the loops as soon as they occur, as described in Chapter 3. When all lines of the system and the machine reactances have been incorporated into the system and axes of buses eliminated, the remaining matrix represents the admittances between the buses of the generators behind the transient reactances as indicated in Ref. 1, p. 79.

The power supplied by the generators can then be obtained by a direct, noniterative method by matrix multiplication as shown in equation 8.20.

$$P = \text{Re}(EI^*) = \text{Re}(EY^*E^*) \tag{8.20}$$

The E is the voltage vector of the generator voltages E_Q; Y^* is the conjugate of the admittance matrix that has been assembled.

The new power values are substituted into equation 8.1 to obtain new voltage angles for modification of the voltage vector of equation 8.20. This is considerably faster than the iterative method.

Some flexibility and loss of detail in the load flow result from use of this method. A decision must be made as to which are the most desirable speed or load flow details. Perhaps both load flow methods should be implemented, and for a particular study the load flow algorithm selected depending on the purpose of the study.

Example 1   The Y-matrix of the system of Fig. 8.9 is to be assembled and only buses 1 and 2 are to be retained.

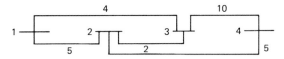

Fig. 8.9.

Since bus 4 is to be eliminated the matrix is assembled considering only the lines connected to bus 4 (see Fig. 8.10).

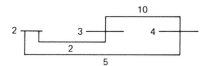

Fig. 8.10.

The bus admittance matrix corresponding to this network is

$$
\begin{array}{c}
 \\
2 \\
3 \\
4
\end{array}
\begin{array}{c}
\begin{array}{ccc}
2 & 3 & 4
\end{array} \\
\left[
\begin{array}{ccc}
7.0 & -2.0 & -5.0 \\
-2.0 & 12.0 & -10.0 \\
-5.0 & -10.0 & 15.0
\end{array}
\right]
\end{array}
$$

The diagonal elements are the sum of the admittances connected to the bus, and the off diagonal elements are the negative of the line admittance between the buses.

The elimination of axis 4 follows the term-by-term modification of equation 3.23 of Chapter 3.

$$Y'_{22} = Y_{22} - Y_{24}Y_{44}^{-1}Y_{42} = 7.0 - \frac{5.0 \times 5.0}{15.0} = 7.0 - 1.6666 = 5.3333$$

$$Y'_{23} = Y_{23} - Y_{24}Y_{44}^{-1}Y_{43} = -2.0 - 3.3333 = -5.3333$$

$$Y'_{33} = Y_{33} - Y_{34}Y_{44}^{-1}Y_{43} = 12.0 - 6.6666 = 5.3333$$

The matrix after elimination of axis 4 is

$$
\begin{array}{cc}
 & \begin{array}{cc} 2 & \quad\quad 3 \end{array} \\
\begin{array}{c} 2 \\ 3 \end{array} &
\left[\begin{array}{cc}
5.3333 & -5.3333 \\
-5.3333 & 5.3333
\end{array} \right]
\end{array}
$$

which represents the network of Fig. 8.11.

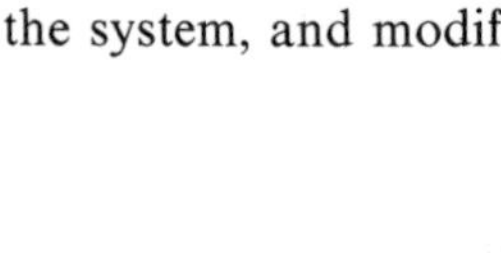

Fig. 8.11.

Add the lines that are connected to bus 3 that have not been included in the system, and modify the matrix for these line additions.

$$
\begin{array}{cc}
 & \begin{array}{ccc} 1 & \quad\quad 2 & \quad\quad 3 \end{array} \\
\begin{array}{c} 1 \\ 2 \\ 3 \end{array} &
\left[\begin{array}{ccc}
4.0 & 0.0 & -4.0 \\
0.0 & 5.3333 & -5.3333 \\
-4.0 & -5.3333 & 9.3333
\end{array} \right]
\end{array}
$$

Elimination of bus 3 is accomplished by use of equation 3.23.

$$Y'_{11} = Y_{11} - Y_{13}Y_{33}^{-1}Y_{31} = 4.0 - \frac{4.0 \times 4.0}{9.3333} = 4.0 - 1.7143 = 2.2857$$

$$Y'_{12} = Y_{12} - Y_{13}Y_{33}^{-1}Y_{32} = 0.0 - \frac{4.0 \times 5.3333}{9.3333} = -2.2857$$

$$Y'_{21} = Y'_{12} = -2.2857$$

$$Y'_{22} = Y_{22} - Y_{23}Y_{33}^{-1}Y_{32} = 5.3333 - \frac{5.3333 \times 5.3333}{9.3333} = 2.2857$$

The network is completed by adding the line of the original system between bus 1 and 2 (see Fig. 8.12).

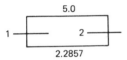

5.0

1 — 2

2.2857   **Fig. 8.12.**

The final matrix is therefore obtained by adding the admittance of this line in the method shown.

$$
\begin{array}{c c}
 & \quad\quad 1 \quad\quad\quad\quad\quad 2 \\
\begin{array}{c} 1 \\ 2 \end{array} &
\begin{bmatrix}
2.2857 + 5.0 & -2.2857 - 5.0 \\
-2.2857 - 5.0 & 2.2857 + 5.0
\end{bmatrix}
\end{array}
$$

TRANSIENT STABILITY STUDY BY COHERENT MACHINES

A method has been reported [15] in which the swing curve of only one machine of each "coherent" group of machines is calculated. The angular position of the other machines of the groups are assumed to swing exactly as does the machine that is representative of the group. This method permits the study of systems that are too large for the computer that is available.

The purpose of a transient stability study is to determine the swing characteristics of the several generators. This method appears to assume the answer of the problem before the study is made. To be sure of the "coherency" of machines a full study should be made for identification of the coherent groups, but the method was devised to eliminate the necessity of making full system studies. Once the full study has been made to establish coherency for a particular system condition, the stability test is completed. The next system condition is a new case and the coherency that has been established probably does not apply.

For example, one of the questions to be answered is whether a particular machine near the border of the system being tested will swing in sympathy with some generator or generators of the test system or will it swing with generators of a neighboring system? Coherency is the problem to be answered and not a premise upon which a study is based.

STEADY-STATE STABILITY

In steady-state stability the admittance between the generator being studied and the infinite system, and the admittance of the effective load connected to ground is required (see p. 355 of Ref. 25, and Fig. 8.13 in this text).

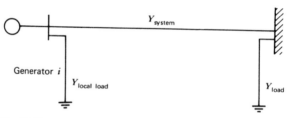

Fig. 8.13.

If the terminal admittance matrix [1] of the generators has been formed, in which the internal reactances of the generators has been included in the network, the admittance, Y_{system}, between any generator and the infinite system can readily be obtained. Each off-diagonal element in row i of the matrix is the negative of the admittance between generator i and another generator of the infinite system. Therefore $\sum_{\substack{j=1 \\ j \neq i}}^{n} Y_{ij}$ is the admittance between generator i and the infinite system.

$$Y_{\text{system}} = -\sum_{\substack{j=1 \\ j \neq i}}^{n} Y_{ij} \tag{8.21}$$

The diagonal elements of the matrix are the sum of all admittances connected to the bus in the equivalent network after all load buses have been eliminated. The admittance to ground is included in the diagonal but does not occur as an off diagonal element. Therefore, the load admittance connected to bus i is obtained by adding all elements in row i.

$$Y_{\text{local load } i} = \sum_{j=1}^{n} Y_{ij} \tag{8.22}$$

The load admittance connected to the infinite bus is the sum of all load admittances except bus i.

$$Y_{\text{load infinite bus}} = \sum_{\substack{j=1 \\ j \neq i}}^{n} Y_{\text{local load } j} \tag{8.23}$$

The three admittances of Fig. 8.13 are therefore all available from the terminal admittance matrix by adding the appropriate elements. The delta of these three admittances is converted to a star connection as shown in Fig. 8.14. It must be remembered that the load impedances of Fig. 8.13 are connected to the buses behind the transient reactance and not to the bus on the high voltage side of the transformer.

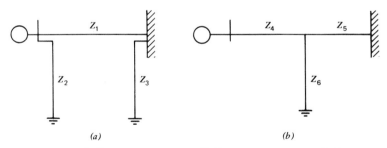

Fig. 8.14. $Z_4 = \dfrac{Z_1 Z_2}{Z_1 + Z_2 + Z_3}$, $Z_5 = \dfrac{Z_1 Z_3}{Z_1 + Z_2 + Z_3}$, $Z_6 = \dfrac{Z_2 Z_3}{Z_1 + Z_2 + Z_3}$.

Included in Z_4 are the reactance of the generator and the transformer. The local load should be connected to point A of Fig. 8.15 which is the high side bus of the transformer.

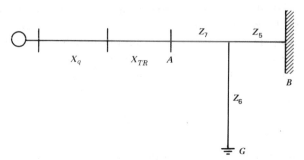

Fig. 8.15.

The star composed of the impedance Z_5, Z_6, Z_7 is converted to a delta.

$$Y_{AB} = \frac{Y_5 Y_7}{Y_5 + Y_6 + Y_7} = \frac{Z_6}{Z_6 Z_7 + Z_5 Z_7 + Z_5 Z_6}$$

$$Y_{AG} = \frac{Z_5}{Z_6 Z_7 + Z_5 Z_7 + Z_5 Z_6}$$

$$Y_{BG} = \frac{Z_7}{Z_6 Z_7 + Z_5 Z_7 + Z_5 Z_6}$$

The circuit that is required for steady-state calculations is shown in Fig. 8.16.

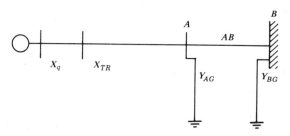

Fig. 8.16.

NETWORK EQUIVALENT WITH DISTRIBUTION FACTORS

In reducing the system to the terminal admittance matrix [1] as indicated in equation 8.19 one can make an interesting extention to the calculation.

Consider the bus admittance of a 4-bus system as given in equation 8.24.

$$
Y = \begin{bmatrix}
Y_{11} & -Y_{12} & -Y_{13} & -Y_{14} \\
-Y_{21} & Y_{22} & -Y_{23} & -Y_{24} \\
-Y_{31} & -Y_{32} & Y_{33} & -Y_{34} \\
-Y_{41} & -Y_{42} & -Y_{43} & Y_{44}
\end{bmatrix}
\tag{8.24}
$$

If instead of the reduction of the matrix by elimination of bus 4 by application of equation 8.19 or equation 3.23, a pivoting operation of the Shipley inversion is made with respect to the pivot element Y_{44} using equations 2.2, 2.3, 2.4, and 2.5. The resulting matrix is given in equation 8.25

$$
Y' = \begin{bmatrix}
Y'_{11} & -Y'_{12} & -Y'_{13} & \dfrac{Y_{14}}{Y_{44}} \\
-Y'_{21} & Y'_{22} & -Y'_{23} & \dfrac{Y_{24}}{Y_{44}} \\
-Y'_{31} & -Y'_{32} & Y'_{33} & \dfrac{Y_{34}}{Y_{44}} \\
\dfrac{Y_{14}}{Y_{44}} & \dfrac{Y_{24}}{Y_{44}} & \dfrac{Y_{34}}{Y_{44}} & -\dfrac{1}{Y_{44}}
\end{bmatrix}
\tag{8.25}
$$

The elements not in the pivot row and column (axes 1, 2, and 3) correspond exactly to the Y-matrix elements resulting from the reduction of the matrix by application of equation 8.19, because equation 2.2 performs the identical operations on the elements as does equation 8.19. In addition row 4 now gives distribution factors to the three remaining buses of the current that originally was injected into bus 4 [18]. Because of symmetry column 4 need not be retained, and the matrix is reduced to that given in equation 8.26.

$$
Y' = \begin{matrix} 1 \\ 2 \\ 3 \\ 4 \end{matrix}
\begin{bmatrix}
Y'_{11} & -Y'_{12} & -Y'_{13} \\
-Y'_{21} & Y'_{22} & -Y'_{23} \\
-Y'_{31} & -Y'_{32} & Y'_{33} \\
D_{4\text{-}1} & D_{4\text{-}2} & D_{4\text{-}3}
\end{bmatrix}
\tag{8.26}
$$

The $D_{4\text{-}1}$, the proportion of the current which formerly was injected into the original network at bus 4, is now injected into the reduced network at bus 1.

Elimination of axis (bus) 3 by a Shipley inversion pivoting on Y'_{33} modifies all elements not in row 3 or column 3 as required by equation 2.2 to give equation 8.27.

$$Y'' = \begin{bmatrix} Y''_{11} & -Y''_{12} & -Y'_{13} \\ -Y''_{21} & Y''_{22} & -Y'_{23} \\ -Y'_{31} & -Y'_{32} & Y'_{33} \\ D'_{4\text{-}1} & D'_{4\text{-}2} & D'_{4\text{-}3} \end{bmatrix} \tag{8.27}$$

Equation 2.3 is used to modify row 3 and column 3 is deleted from the matrix to give equation 8.28.

$$Y'' = \begin{bmatrix} Y''_{11} & -Y''_{12} \\ -Y''_{21} & Y''_{22} \\ D_{3\text{-}1} & D_{3\text{-}2} \\ D'_{4\text{-}1} & D'_{4\text{-}2} \end{bmatrix} \tag{8.28}$$

The Y elements in the matrix correspond to the admittance matrix of the reduced system, and the D elements are distribution factors. Note that $D'_{4\text{-}1}$ and $D'_{4\text{-}2}$ includes the redistribution of the current from 4-3 that has changed the distribution of $D_{4\text{-}1}$ and $D_{4\text{-}2}$.

Example 2   The complete matrix corresponding to the network of Fig. 8.9 is used because the assembly algorithm of the matrix would unnecessarily complicate the example and might obscure the distribution of current calculation.

$$
Y = \begin{array}{c} \\ 1 \\ 2 \\ 3 \\ 4 \end{array}
\begin{array}{cccc}
1 & 2 & 3 & 4 \\
\end{array}
\begin{bmatrix}
9.0 & -5.0 & -4.0 & 0.0 \\
-5.0 & 12.0 & -2.0 & -5.0 \\
-4.0 & -2.0 & 16.0 & -10.0 \\
0.0 & -5.0 & -10.0 & 15.0
\end{bmatrix} \tag{8.29}
$$

Perform a Shipley inversion pivoting operation on element Y_{44} and delete column 4 to give equation 8.30.

$$Y' = \begin{array}{c} \\ 1 \\ 2 \\ 3 \\ 4 \end{array} \begin{array}{ccc} 1 & 2 & 3 \\ \left[\begin{array}{ccc} 9.0 & -5.0 & -4.0 \\ -5.0 & 10.3333 & -5.3333 \\ -4.0 & -5.3333 & 9.3333 \\ 0 & 0.3333 & 0.6666 \end{array}\right] \end{array} \qquad (8.30)$$

The distribution factors indicate that one-third of the current that was injected into bus 4 has now been transferred to bus 2, two-thirds has been transferred to bus 3, and none to bus 1.

Perform a pivoting operation on element Y_{33} to give the matrix 8.31.

$$Y'' = \begin{array}{c} \\ 1 \\ 2 \\ 3 \\ 4 \end{array} \begin{array}{cc} 1 & 2 \\ \left[\begin{array}{cc} 7.2857 & -7.2857 \\ -7.2857 & 7.2857 \\ 0.4300 & 0.5700 \\ 0.2860 & 0.7140 \end{array}\right] \end{array} \qquad (8.31)$$

Observe that 0.333 of the current originally injected into bus 4 arrived at bus 2 when bus 4 was eliminated. It has now increased to 0.714 because of the redistribution of the bus 4 current that was originally distributed to bus 3.

Example 3 To show the effect of loads represented as ground ties consider the same network but with an admittance to ground connected to bus 4 (see Fig. 8.17).

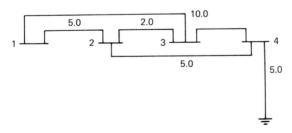

Fig. 8.17.

Without giving the details the corresponding matrix is given in equation 8.32.

$$Y = \begin{bmatrix} 9.0 & -5.0 & -4.0 & 0.0 \\ -5.0 & 12.0 & -2.0 & -5.0 \\ -4.0 & -2.0 & 16.0 & -10.0 \\ 0.0 & -5.0 & -10.0 & 20.0 \end{bmatrix} \qquad (8.32)$$

The first pivoting operation and deletion gives equation 8.33.

$$Y' = \begin{matrix} & 1 & 2 & 3 \\ 1 \\ 2 \\ 3 \\ 4 \end{matrix} \begin{bmatrix} 9.0 & -5.00 & -4.00 \\ -5.0 & 10.75 & -4.50 \\ -4.0 & -4.50 & 11.00 \\ 0.0 & 0.25 & 0.50 \end{bmatrix} \qquad (8.33)$$

Note that the distribution of the current injection into bus 4 no longer adds up to unity, because 25% of the current injected into bus 4 in the original network supplied local load and never appeared on the transmission system. The pivoting on element Y_{33} is left for an exercise for the student.

Whether to include load admittances is a matter of choice and the purpose of the equivalent in the study to be made. For a very detailed discussion that includes auxiliary control signals for the voltage regulator, governor action for both steam, and hydro turbines and relay action to be included in the study, the reader is referred to the excellent paper by C. C. Young [26].

References

1. **E. W. Kimbark,** *Power System Stability*, Vol. 1, Wiley, 1948.
2. **S. D. Crary,** *Power System Stability*, Vol. 2, General Electric Co., Schenectady, N.Y., 1947.
3. **V. A. Venikov,** *Transient Phenomina in Electric Power Systems*, Pergamon Press, 1965.
4. **W. E. Milne,** *Numerical Solutions of Differential Equations*, Wiley, 1953.

5. **L. Collatz,** *The Numerical Treatment of Differential Equations*, Springer-Verlag OHG, 1966.

6. **O. Elgerd,** *Control Systems Theory*, McGraw-Hill, 1967.

7. **M. S. Dyrkacz, C. C. Young, and F. J. Maginniss,** A digital transient stability program including the effects of regulator, exciter and governor response, *Trans. AIEE*, Vol. 79 Part 3, (1960), pp. 1245–1257.

8. **D. W. Olive,** New techniques for the calculation of dynamic stability, *Trans. AIEE PA&S*, Vol. 85, (1960), pp. 767–777.

9. **M. H. Kent et al.,** Dynamic modeling of loads in stability studies, *IEEE Trans. PA&S*, Vol. 88, (May 1969), pp. 756–763.

10. **C. C. Young,** Equipment and system modeling for large-scale stability studies, *IEEE PICA, 1971 Conf. Proc. TP V-B*, pp. 163–172.

11. **H. E. Lokay and R. L. Bolger,** Effect of turbine-generator representation in system stability studies, *IEEE Trans. PA&S*, (1965), p. 933.

12. **E. W. Kimbark,** discussion of Ref. 8.

13. *Tutorial Course of IEEE*, Publication 70M62-PWR, E. W. Kimbark, C. C. Young, F. P. deMello, F. G. McCrackin, W. R. Schmus, G. W. Stagg, and W. F. Tinney.

14. **E. W. Kimbark,** discussion of Ref. 11.

15. **A. Chang and M. M. Adibi,** Power system dynamic equivalent, *Proc. Power Industry Computer Applications Conf.*, Denver, May 1969, pp. 174–183.

16. **L. J. Rindt, R. W. Long, and R. T. Byerly,** Transient stability, Part 3, *Trans. AIEE PA&S*, Vol. 73, (1959), p. 1673.

17. **P. L. Dandeno and K. Prabhashankar,** A non-iterative transient stability program including the effects of variable load-voltage characteristics, *IEEE Trans.*, Paper 773-891-6.

18. **H. E. Brown, R. B. Shipley, D. Coleman, and R. E. Neid, Jr.,** A study of stability equivalents, *IEEE Trans. PA&S*, Vol. 88, (March 1969), p. 200.

19. **W. F. Tinney and J. W. Walker,** Direct solution of sparce network equations by optimally ordered triangular factorization, *Proc. IEEE*, Vol. 55, (November 1967), pp. 1801–1809.

20. **W. F. Tinney and C. E. Hart,** Power flow solutions by Newton's method, *IEEE Trans. PA&S*, Vol. 86, (November 1967), p. 1449.

21. **C. Concordia,** Steady state stability of synchronous machines as affected by voltage regulator characteristics, *Trans. AIEE*, Vol. 63, (1944), pp. 215–220.

22. *Report on Answers to Questionaire on Additional Signals in Excitation Systems*, Working Group 03 of Study Committee No. 32, Cigre, Electra No. 26, January 1973.

23. IEEE Committee Report, Computer representation of excitation systems, *IEEE Trans. PA&S*, Vol. 87, (June 1968), pp. 1460–1464.

24. IEEE Committee Report, Excitation systems dynamic characteristics, *IEEE Trans.*, Paper T 72 590-8.

25. **J. R. Neuenswander,** *Modern Power Systems*, International Textbook, 1971.

26. **C. C. Young,** Equipment and system modeling for large-scale stability studies, *IEEE, PICA Proc.*, Boston, 1971, p. 163.

9

Eigenvalues, Eigenvectors; Linear Programming and Optimization

A single book on network analysis must necessarily be limited. This book presents those techniques that are useful in solving large power system transmission network problems. Any technique for the solution of power network problems that have not been used by the author during the past 20 years, for example, loop equations, are not considered to be used nor useful and are therefore not discussed. Through research effort in the area of network analysis, a known technique but unused because of some technical difficulty, may suddenly become the method of choice. The risk is great that an omitted technique may become important by the time the book is printed. This book therefore presents the state of the art at the time of writing.

Important areas have been ignored either because the material is completely outside the author's experience or because it is so voluminous as to require a larger treatment. Transient voltage analysis is an excellent example, since it fits both catagories. State estimation techniques and several other topics have been omitted because of lack of space.

Three additional topics are discussed briefly: eigenvalues and eigenvectors, linear programming, and optimization techniques. Eigenvalues and eigenvectors were selected because of the considerable gap between the technical level of the papers on the subject and the general understanding of the material. The author tries to bridge this gap and indicate areas where the method holds great promise. Source materials are listed for further research and study by individuals interested in the area.

A discussion of linear programming indicates the great potential of this technique. The outline of this method permits writing a program for a small problem. For large problems, a copy of a very sophisticated program should be obtained from a computer service bureau or from a computer users' group. The use of this technique will certainly increase, because of the many possible applications in the power industry. Dynamic programming and integer programming have been omitted for lack of space.

The load flow, short-circuit, and transient stability programs have commonly been referred to as system planning programs. Actually the planning is done by an engineer or group of engineers, and these "system planning programs" merely determine whether the proposed plan for future system expansion is feasible. Optimization techniques can (and must be) used in future system planning. The optimum allocation of capacitors for supplying vars to the system, the location of substations to minimize line investment, and many other applications are possible. Thus the computer can solve these important problems whereas manual findings can not.

EIGENVALUES AND EIGENVECTORS

Assume that a set of n linearly independent equations in n unknowns is given. As a matter of convenience and for easier understanding, the following illustrations generally are limited to two equations in two unknowns. Consider the set of equations 9.1

$$a_{11}x_1 + a_{12}x_2 = y_1$$

$$a_{21}x_1 + a_{22}x_2 = y_2 \qquad (9.1)$$

These equations can be expressed in matrix notation by equation 9.2.

$$\begin{bmatrix} a_{11} & a_{12} \\ a_{21} & a_{22} \end{bmatrix} \begin{bmatrix} x_1 \\ x_2 \end{bmatrix} = \begin{bmatrix} y_1 \\ y_2 \end{bmatrix} \qquad (9.2)$$

The notation can be further simplified to read

$$AX = Y \qquad (9.3)$$

In the discussions to follow A will be a square n by n matrix; X and Y will both be nth-order vectors.

The problem is to find a vector X (called eigenvector, characteristic vector, proper vector, or modal column [1–3]) which will be transformed by the matrix A into a vector Y, whose coordinates are proportional to those of X, the original vector, and therefore has the same direction as X in the vector space. That is, find a vector X which satisfies the equation 9.4 where λ (called the eigenvalue, characteristic value, characteristic number, or proper number) is a scalar, real, or complex that must also be determined.

$$AX = \lambda X \tag{9.4}$$

Since λ is a scalar, λX can be replaced by means of the identity 9.5 where I is the identity matrix and X is a vector.

$$\lambda X \equiv \lambda I X \tag{9.5}$$

Substitution of equation 9.5 into equation 9.4 gives equation 9.6.

$$AX = \lambda I X \tag{9.6}$$

Collecting terms and making use of the distributive law of matrix multiplication give equation 9.7

$$[A - \lambda I][X] = 0 \tag{9.7}$$

The matrix $[A - \lambda I]$ is called the characteristic matrix of the matrix A. It represents the coefficients of a set of homogeneous simultaneous linear equations which will have a solution other than the trivial solution $X \equiv 0$, only if the determinant of the matrix $[A - \lambda I]$ is identically zero.

Evaluation of $\det|A - \lambda I|$ results in an nth degree polynomial in λ, which is called the characteristic equation. The roots of the polynomial are the eigenvalues of the matrix A.

PHYSICAL INTERPRETATION OF THE PROBLEM

Every point in the n-dimensional space can be considered the tip of a vector emanating from an arbitrarily selected origin of the coordinate system of the vector space. A two-dimensional space is used for illustration purposes but it must be remembered that in most engineering problems, the number of variables is very large [5].

A matrix A transforms every point in the space (every vector) into another vector in the space. For example, the matrix

$$A = \begin{bmatrix} 1 & 3 \\ 2 & 2 \end{bmatrix}$$

transforms the vector

$$X = \begin{bmatrix} 1 \\ 0 \end{bmatrix}$$

into the vector

$$Y = \begin{bmatrix} 1 \\ 2 \end{bmatrix}$$

according to equation 9.2. See equation 9.8.

$$AX = \begin{bmatrix} 1 & 3 \\ 2 & 2 \end{bmatrix} \begin{bmatrix} 1 \\ 0 \end{bmatrix} = \begin{bmatrix} 1 \\ 2 \end{bmatrix} \tag{9.8}$$

This transformation is illustrated in Fig. 9.1. Notice that the two vectors have different directions in the vector space. In eigenvalue, eigenvector problems one must find the vectors X which will be transformed by the matrix A into vectors Y that have the same direction in space as the original vectors.

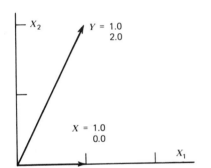

Fig. 9.1.

Applying equation 9.7 to the example, gives equation 9.9

$$[A - I\lambda][X] = \left[\begin{bmatrix} 1 & 3 \\ 2 & 2 \end{bmatrix} - \begin{bmatrix} \lambda & 0 \\ 0 & \lambda \end{bmatrix} \right] \begin{bmatrix} X_1 \\ X_2 \end{bmatrix} = \begin{bmatrix} 1-\lambda & 3 \\ 2 & 2-\lambda \end{bmatrix} \begin{bmatrix} X_1 \\ X_2 \end{bmatrix} = 0$$

$$\tag{9.9}$$

Expand the determinant of the characteristic matrix to obtain the characteristic polynomial, which must be equal to zero if the vectors are to have other than the trivial solution.

$$\det \begin{vmatrix} 1-\lambda & 3 \\ 2 & 2-\lambda \end{vmatrix} = -4-3\lambda+\lambda^2=0 \qquad (9.10)$$

The roots of this equation are the two eigenvalues $\lambda_1=4$ and $\lambda_2=-1$. Substitution of the value $\lambda_1=4$ back into equation 9.9 gives equation 9.11

$$\begin{bmatrix} 1-\lambda & 3 \\ 2 & 2-\lambda \end{bmatrix}\begin{bmatrix} X_1 \\ X_2 \end{bmatrix} = \begin{bmatrix} -3 & 3 \\ 2 & -2 \end{bmatrix}\begin{bmatrix} X_1 \\ X_2 \end{bmatrix} = 0 \qquad (9.11)$$

This set of simultaneous homogeneous linear equations does not have a unique solution. However, when one unknown has been assigned a value, the other unknown can be determined. For example if $X_1=1$ then $X_2=1$. The matrix A transforms the vector

$$X = \begin{bmatrix} 1 \\ 1 \end{bmatrix}$$

into the vector

$$Y = \begin{bmatrix} 4 \\ 4 \end{bmatrix}$$

which has the same direction in space as X. (The magnitude of the vector has been changed but this is not important.)

The reader must verify that the second eigenvalue $\lambda_2=-1$ results in a vector which is transformed by the matrix A into a vector with the same direction in space. Note that the two vectors will have the same direction in space, but will be oppositely oriented.

A 3 by 3 matrix is used to further illustrate the techniques used in eigenvalue problems for a better understanding. Consider

$$A = \begin{bmatrix} 8 & -8 & -2 \\ 4 & -3 & -2 \\ 3 & -4 & 1 \end{bmatrix} \qquad (9.12)$$

The determinant of the characteristic matrix must be expanded and set equal to zero.

$$\det \begin{vmatrix} 8-\lambda & -8 & -2 \\ 4 & -3-\lambda & -2 \\ 3 & -4 & 1-\lambda \end{vmatrix} = 0 \qquad (9.13)$$

Expansion of the determinant gives the characteristic polynomial 9.14.

$$-\lambda^3 + 6\lambda^2 - 11\lambda + 6 = 0 \qquad (9.14)$$

The roots of this polynomial must be determined; $\lambda_1 = 1$, $\lambda_2 = 2$, and $\lambda_3 = 3$. Substitution of $\lambda_1 = 1$ back into equation 9.13 defines the three simultaneous equations of 9.15.

$$\begin{bmatrix} 7 & -8 & -2 \\ 4 & -4 & -2 \\ 3 & -4 & 0 \end{bmatrix} \begin{bmatrix} X_1 \\ X_2 \\ X_3 \end{bmatrix} = 0 \qquad (9.15)$$

A solution of equation 9.15 is obtained by arbitrarily setting $X_3 = 2$.

$$\begin{bmatrix} X_1 \\ X_2 \\ X_3 \end{bmatrix} = \begin{bmatrix} 4 \\ 3 \\ 2 \end{bmatrix} \qquad (9.16)$$

Determination of the two vectors corresponding to the other two eigenvalues is left for the reader. These two examples illustrate the technique, but if the order of the matrix A of equation 9.7 is large, the evaluation of the determinant of the matrix $[A - \lambda I]$ to obtain the characteristic equation and subsequently to determine the λ roots and the corresponding eigenvectors becomes a great computational burden.

Fortunately in many engineering problems only the largest (smallest) eigenvalue is of interest. In other cases one must know if an eigenvalue is close to a particular numerical value. In still other instances, one must only determine if the real part of all eigenvalues is negative. These techniques are examined briefly. These problems are usually much simpler than the determination of the entire set of eigenvalues; however, if there are equal roots or roots that have nearly equal values, some of the techniques have difficulty. Before describing the methods of computation in these specialized techniques, we discuss the inherent reason for difficulty in certain cases.

QUADRATIC FORMS

Associated with the square matrix A is a binary form in which A is premultiplied by a row vector Y and postmultiplied by a column vector X as shown in equation 9.17

$$B(y,x)=[y_1 y_2 \cdots y_n][A]\begin{bmatrix} X_1 \\ X_2 \\ \vdots \\ X_n \end{bmatrix}=0 \qquad (9.17)$$

For example, the binary form related to the matrix used in the earlier example is given in equation 9.18.

$$B(y,x)=[y_1 y_2]\begin{bmatrix} 1 & 3 \\ 2 & 2 \end{bmatrix}\begin{bmatrix} x_1 \\ x_2 \end{bmatrix}=x_1 y_1+3x_2 y_1+2x_1 y_2+2x_2 y_2 \quad (9.18)$$

The coefficient matrices in many engineering problems are symmetric. (The off-diagonal elements satisfy the relationship $a_{ij}=a_{ji}$.) As a matter of convenience, most of the material that follows considers symmetric matrices, although many of the techniques apply equally as well to matrices that are unsymmetric.

If A is symmetric and $y_i=x_i$, then the binary form becomes quadratic as shown in equation 9.19.

$$Q(X,X)=[X_1 X_2 \cdots X_n][A]\begin{bmatrix} X_1 \\ X_2 \\ \vdots \\ X_n \end{bmatrix}=k \qquad (9.19)$$

For example if

$$A = \begin{bmatrix} 3 & 1 \\ 1 & 3 \end{bmatrix}$$

then the associated quadratic form is given in equation 9.20.

$$Q(X,X) = [X_1 X_2] \begin{bmatrix} 3 & 1 \\ 1 & 3 \end{bmatrix} \begin{bmatrix} X_1 \\ X_2 \end{bmatrix} = 3X_1^2 + 2X_1X_2 + 3X_2^2 = k \quad (9.20)$$

Every point in the two-dimensional space belongs to some member of this infinite family of conic sections obtained by assigning the full range of values $0 < k < \infty$ in equation 9.20. The vectors that correspond to a particular value of k describe a "level curve," a curve of constant field intensity. It is shown in textbooks on vector analysis [4] that the matrix A transforms a vector X into another vector that points in the direction of the gradient from the point X on the surface, or the direction in which the field intensity changes the most rapidly for an incremental movement of the vector from the given point on the surface. Thus equation 9.21 determines the gradient vector Y at the point X on the surface defined by equation 9.22

$$AX = Y \quad\quad\quad\quad (9.21)$$

$$X^T A X = K \quad\quad\quad\quad (9.22)$$

For example, assign the value of 12 to K in equation 9.20 to obtain equation 9.23

$$3X_1^2 + 2X_1X_2 + 3X_2^2 = 12 \quad\quad\quad\quad (9.23)$$

The level curve of equation 9.23 is shown in Fig. 9.2. Four vectors, X and the corresponding gradient vectors Y are also shown in Table 9.1. The two vectors X_3 and X_4 satisfy the condition of equation 9.4 and are therefore eigenvectors of the matrix A. These two vectors must be normal to the curve, since they have the same direction as the gradient vectors Y_3 and Y_4. The radius vectors of a conic that are at right angles to the surface are its major and minor axes. Therefore, the major and minor axes of the quadratic form are the eigenvectors of the matrix A.

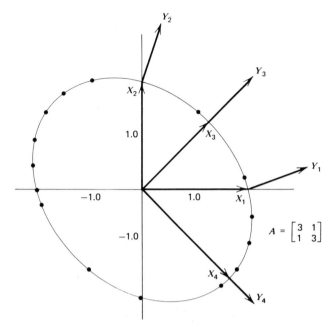

Fig. 9.2. Level curve $3X_1^2 + 2X_1X_2 + 3X_2^2 = 12$.

$$A = \begin{bmatrix} 3 & 1 \\ 1 & 3 \end{bmatrix}$$

Table 9.1

Vector	X	Y
X_1	$\begin{bmatrix} 2.0 \\ 0.0 \end{bmatrix}$	$\begin{bmatrix} 6.0 \\ 2.0 \end{bmatrix}$
X_2	$\begin{bmatrix} 0.0 \\ 2.0 \end{bmatrix}$	$\begin{bmatrix} 2.0 \\ 6.0 \end{bmatrix}$
X_3	$\begin{bmatrix} \sqrt{1.5} \\ \sqrt{1.5} \end{bmatrix}$	$4\begin{bmatrix} \sqrt{1.5} \\ \sqrt{1.5} \end{bmatrix}$
X_4	$\begin{bmatrix} \sqrt{3.0} \\ -\sqrt{3.0} \end{bmatrix}$	$2\begin{bmatrix} \sqrt{3.0} \\ -\sqrt{3.0} \end{bmatrix}$

It should also be noted that the vectors Y_1 and Y_2 both point in a direction closer to the direction of the eigenvector X_3 than do the corresponding vectors X_1 and X_2. This suggests that an iterative technique might very well be used to find an eigenvector.

Arbitrarily assume an original vector, say

$$X_1 = \begin{bmatrix} 2 \\ 0 \end{bmatrix}$$

and compute $AX_1 = Y_1$.

$$Y_1 = \begin{bmatrix} 6.0 \\ 2.0 \end{bmatrix} = 2 \begin{bmatrix} 3.0 \\ 1.0 \end{bmatrix}$$

In every case the new vector will be divided by the last element of the vector to prevent difficulties of scaling in the iterative process. Repeat the steps using $X_2 = Y_1$ as the new estimate of the eigenvector.

$$AY_1 = \begin{bmatrix} 3 & 1 \\ 1 & 3 \end{bmatrix} \begin{bmatrix} 3.0 \\ 1.0 \end{bmatrix} = \begin{bmatrix} 10.0 \\ 6.0 \end{bmatrix} = 6.0 \begin{bmatrix} 1.66666 \\ 1.00000 \end{bmatrix} = \lambda_2 Y_2$$

In the next iteration use Y_2 as the estimate of the eigenvector.

$$AY_2 = \begin{bmatrix} 3 & 1 \\ 1 & 3 \end{bmatrix} \begin{bmatrix} 1.66666 \\ 1.00000 \end{bmatrix} = \begin{bmatrix} 6.0000 \\ 4.6666 \end{bmatrix} = 4.6666 \begin{bmatrix} 1.2857 \\ 1.0000 \end{bmatrix}$$

After 10 iterations the eigenvector and eigenvalue are approximated by

$$\begin{bmatrix} 3 & 1 \\ 1 & 3 \end{bmatrix} \begin{bmatrix} 1.000976 \\ 1.000000 \end{bmatrix} = \begin{bmatrix} 4.002930 \\ 4.0009767 \end{bmatrix} = 4.0008177 \begin{bmatrix} 1.00048 \\ 1.00098 \end{bmatrix}$$

The process indicates that $\lambda = 4.0$ and the corresponding eigenvector is

$$\begin{bmatrix} 1.0 \\ 1.0 \end{bmatrix}$$

Solving for the eigenvalue by the earlier method gives the characteristic equation

$$\begin{bmatrix} 3-\lambda & 1 \\ 1 & 3-\lambda \end{bmatrix} = 8 - 6\lambda + \lambda^2 = 0$$

The roots are 4.0 and 2.0. The process has yielded the largest eigenvalue and the corresponding eigenvector.

The greater the difference between the absolute values of the largest and second largest eigenvalue, the faster will be the convergence of the process that has just been described. It is suggested as an exercise that the student obtain the largest eigenvalue of the matrix $A = \begin{bmatrix} 1 & 4 \\ 4 & 7 \end{bmatrix}$ using five iterations. The convergence is much faster than that in the previous example.

Conversely, if the two largest eigenvalues are nearly equal in magnitude, the iterative process will be extremely slow. If the absolute values of two eigenvalues differ by a small amount (Fig. 9.2), the conic is nearly circular, and it becomes more difficult to determine which is the smallest radius vector which corresponds to the largest eigenvalue. The eigenvalue thus determined will be much more accurate than the eigenvector. If only the eigenvalue is required, the process is moderately successful in this difficult case.

SHIFTING

In a particular problem it might happen that the eigenvalue with the smallest absolute value is required. By a process of shifting, the smallest eigenvalue can be made to be the eigenvalue with the largest absolute value. Subtract KIX from both sides of equation 9.4 to obtain equation 9.24. This is equivalent to subtraction of a constant K from every diagonal element of the matrix A, which is equivalent to a shift in axis of K

$$[A - KI]X = [\lambda I - KI]X \tag{9.24}$$

Define $|A - KI| = B$ and $|\lambda I - KI| = \mu$. This equation becomes equation 9.25 which is easily recognized to be identical to equation 9.4 using new symbols.

$$BX = \mu X \tag{9.25}$$

Equation 9.25 can be solved for μ by the iterative method, and the eigenvalue that is obtained can be shifted back by the amount K to give the smallest λ value.

It was found earlier that the largest λ of the matrix $A = \begin{bmatrix} 1 & 4 \\ 4 & 7 \end{bmatrix}$ was

equal to 9.0. To obtain the other eigenvalue a shift of 9 to the left would reduce the corresponding λ, by 9 to 0.0.

The eigenvalue that had the smallest absolute value now becomes the eigenvalue with the largest absolute value. It can be solved for by the iterative process and when it has been obtained it can be shifted back to its correct value.

$$[A - 9I] = \begin{bmatrix} 1-9 & 4 \\ 4 & 7-9 \end{bmatrix} = \begin{bmatrix} -8 & 4 \\ 4 & -2 \end{bmatrix} = B$$

The first iteration using the same original estimate for

$$X = \begin{bmatrix} 2.0 \\ 0.0 \end{bmatrix}$$

gives

$$BX = \begin{bmatrix} -8 & 4 \\ 4 & -2 \end{bmatrix} \begin{bmatrix} 2 \\ 0 \end{bmatrix} = \begin{bmatrix} -16 \\ 8 \end{bmatrix} = 8 \begin{bmatrix} -2 \\ 1 \end{bmatrix}$$

The second iteration is

$$\begin{bmatrix} -8 & 4 \\ 4 & -2 \end{bmatrix} \begin{bmatrix} -2.0 \\ 1.0 \end{bmatrix} = \begin{bmatrix} 20.0 \\ -10.0 \end{bmatrix} = -10 \begin{bmatrix} -2.0 \\ 1.0 \end{bmatrix}$$

The third iteration is

$$\begin{bmatrix} -8 & 4 \\ 4 & -2 \end{bmatrix} \begin{bmatrix} -2.0 \\ 1.0 \end{bmatrix} = \begin{bmatrix} 20.0 \\ -10.0 \end{bmatrix} = -10 \begin{bmatrix} -2.00 \\ 1.00 \end{bmatrix}$$

Since iteration 3 repeats the value of iteration 2, the iteration is stopped

$$\mu_2 = -10.0$$

$$\lambda_2 = \mu_2 + K = -10.00 + 9.0 = -1.0000 \tag{9.26}$$

In practical problems often the small eigenvalues are closely bunched, and the evaluation by the iterative technique is difficult, if not unsuccessful, in spite of the shifting method.

An important transformation of the symmetric matrix A that is used very frequently in engineering applications is the similarity transformation of equation 9.27

$$C = B^{-1}AB \qquad (9.27)$$

It is called the similarity transformation because the eigenvalues of the matrix C are identical to the eigenvalues of the symmetric matrix A.

The characteristic matrix of matrix C is obtained by use of equation 9.7

$$[C - \lambda I] = [B^{-1}AB - \lambda I] \qquad (9.28)$$

Since $B^{-1}B = I$, the last term of equation 9.28 would be unchanged if it is premultiplied by $B^{-1}B$.

$$C - \lambda I = B^{-1}BB^{-1}AB - B^{-1}B\lambda I = B^{-1}[AB - B\lambda I] \qquad (9.29)$$

The order in which the identity matrix enters a multiplication need not be maintained, and since λ is a scalar, $B\lambda I$ can be written λIB.

This alteration in equation 9.29 produces equation 9.30

$$C - \lambda I = B^{-1}[A - \lambda I]B \qquad (9.30)$$

The characteristic equation of this equation is obtained by evaluating the determinant on both sides of the equal sign.

$$\det|C - \lambda I| = \det|B^{-1}| \det|A - \lambda I| \det|B| \qquad (9.31)$$

Remembering that determinants are scalars and the order of the terms in a product need not be preserved, and furthermore that $\det|B^{-1}| \det|B| = 1.0$, then one can reduce equation 9.31 to equation 9.32 which has equal roots.

$$\det|C - \lambda I| = \det|A - \lambda I| \qquad (9.32)$$

Therefore the characteristic equation of C and A are equal, and the eigenvalues of C are identical to those of A.

Exercise

Given

$$A = \begin{bmatrix} 3 & 1 \\ 1 & 3 \end{bmatrix}, \quad B = \begin{bmatrix} 2 & 1 \\ 7 & 4 \end{bmatrix}, \quad \text{and} \quad B^{-1} = \begin{bmatrix} 4 & -1 \\ -7 & 2 \end{bmatrix}$$

The characteristic matrix is

$$[A - \lambda I] = \begin{bmatrix} 3-\lambda & 1 \\ 1 & 3-\lambda \end{bmatrix}$$

The characteristic equation is

$$p(\lambda) = \det \begin{bmatrix} 3-\lambda & 1 \\ 1 & 3-\lambda \end{bmatrix} = 8 - 6\lambda + \lambda^2$$

The eigenvalues are $\lambda_1 = 2$ and $\lambda_2 = 4$, and the corresponding eigenvectors are $\begin{bmatrix} 1 \\ -1 \end{bmatrix}$ and $\begin{bmatrix} 1 \\ 1 \end{bmatrix}$.

An exercise for the reader is to verify that $C = B^{-1}AB$ has the same characteristic equation as matrix A. Very powerful methods have been developed for obtaining all of the eigenvalues of a matrix by means of similarity transformations [6, 7].

THE LR-TRANSFORMATION

This brief treatment of eigenvalues was not intended to supply the mathematical background necessary for a complete understanding of the material. From this point on the methods are presented without justification. If the reader is interested, he should consult the considerable source material listed in Ref. 6.

A symmetric matrix A_1 can be decomposed into a lower left matrix L_1, whose diagonal elements are all unity and an upper right matrix is R_1, such that

$$A_1 = L_1 R_1 \tag{9.33}$$

If the matrices L_1 and R_1 are reversed in the product, the resulting matrix will have the same eigenvalues as A_1.

$$A_2 = R_1 L_1 \tag{9.34}$$

The lower left off-diagonal elements of A_2 will usually be smaller than those of the matrix A_1, and the diagonal elements will approach the value of the eigenvalues. The matrix A_2 is decomposed into L_2 and R_2, and A_3 is formed by obtaining the product of L_2 and R_2 in the reverse order; the lower left off-diagonal elements are further reduced. As the process is repeated many times, the diagonal terms of A_n approach the eigenvalues of the matrix A_1 to any desired precision.

The decomposition of the matrix A_1 into L_1 and R_1 is effectively accomplished by performing a Gaussian elimination and at the same time keeping account of the steps that are necessary for a back substitution. Using the example from Section 4.3 of Ref. 6 we show the steps in detail.

$$A_1 = \begin{bmatrix} 5 & 4 & 1 & 1 \\ 4 & 5 & 1 & 1 \\ 1 & 1 & 4 & 2 \\ 1 & 1 & 2 & 4 \end{bmatrix} \tag{9.35}$$

Multiplying the first row by $\frac{4}{5}$ and subtracting from row 2, and multiplying the first row by $\frac{1}{5}$ and subtracting from rows 3 and 4 gives equation 9.36.

$$A_2 = \begin{bmatrix} 5 & 4.0 & 1.0 & 1.0 \\ 0 & 1.8 & 0.2 & 0.2 \\ 0 & 0.2 & 3.8 & 1.8 \\ 0 & 0.2 & 1.8 & 3.8 \end{bmatrix} \tag{9.36}$$

Similar steps that have been performed to modify matrix A_1 are used to modify the identity matrix B_1.

$$B_1 = \begin{bmatrix} 1 & 0 & 0 & 0 \\ 0 & 1 & 0 & 0 \\ 0 & 0 & 1 & 0 \\ 0 & 0 & 0 & 1 \end{bmatrix} \tag{9.37}$$

The first row is multiplied by $\frac{4}{5}$ and *added* to the second row, and multiplied by $\frac{1}{5}$ and *added* to the third and fourth rows

$$B_1' = \begin{bmatrix} 1.0 & 0 & 0 & 0 \\ 0.8 & 1 & 0 & 0 \\ 0.2 & 0 & 1 & 0 \\ 0.2 & 0 & 0 & 1 \end{bmatrix} \tag{9.38}$$

The elements of the second row begining at the diagonal element of matrix 9.36 are multiplied by $0.2/1.8$ and subtracted from rows 3 and 4.

$$A_1'' = \begin{bmatrix} 5 & 4.0 & 0 & 1 \\ 0 & 1.8 & 0.200000 & 0.200000 \\ 0 & 0 & 3.777778 & 1.777778 \\ 0 & 0 & 1.777778 & 3.777778 \end{bmatrix} \tag{9.39}$$

Performing the multiplication and *addition* operations on matrix B_1' gives equation 9.40

$$B_1'' = \begin{bmatrix} 1.0 & 0 & 0 & 0 \\ 0.8 & 1.000000 & 0 & 0 \\ 0.2 & 0.1111111 & 1 & 0 \\ 0.2 & 0.1111111 & 0 & 1 \end{bmatrix} \tag{9.40}$$

Modifying the matrix A_1'' by multiplying the elements in the third row, beginning at the diagonal element by 1.777778/3.777778, and subtracting from the fourth row give equation 9.41

$$A_1'' = R_1 = \begin{bmatrix} 5 & 4.0 & 1 & 1 \\ 0 & 1.8 & 0.200000 & 0.200000 \\ 0 & 0 & 3.777778 & 1.777778 \\ 0 & 0 & 0 & 2.941176 \end{bmatrix} \tag{9.41}$$

The corresponding modification is also made to matrix B_1''.

$$B_1''' = L_1 = \begin{bmatrix} 1.0 & 0 & 0 & 0 \\ 0.8 & 1.000000 & 0 & 0 \\ 0.2 & 0.1111111 & 1.00000 & 0 \\ 0.2 & 0.1111111 & 0.470588 & 1 \end{bmatrix} \tag{9.42}$$

The reader may wish to verify that $L_1 R_1 = A_1$. Form the matrix A_2 by reversing the two factors L_1 and R_1.

$$A_2 = R_1 L_1 = \begin{bmatrix} 5 & 4.0 & 1.000000 & 1.000000 \\ 0 & 1.8 & 0.200000 & 0.200000 \\ 0 & 0 & 3.777778 & 1.777778 \\ 0 & 0 & 0 & 2.941176 \end{bmatrix} \begin{bmatrix} 1.0 & 0 & 0 & 0 \\ 0.8 & 1.0900000 & 0 & 0 \\ 0.2 & 0.1111111 & 1.000000 & 0 \\ 0.2 & 0.1111111 & 0.470588 & 1 \end{bmatrix}$$

$$A_2 = \begin{bmatrix} 8.6000000000 & 4.2222220 & 1.470588 & 1.000000 \\ 1.5200000000 & 1.8444440 & 0.294118 & 0.200000 \\ 1.1111111111 & 0.6172840 & 4.614379 & 1.777778 \\ 0.5882350000 & 0.3267973 & 1.384083 & 2.941176 \end{bmatrix} \tag{9.43}$$

To illustrate a method that has quadratic convergence compared to the linear convergence of the LR transformation method, this process is carried one more step without giving the computational details. Decomposition of A_2 into L_2 and R_2 gives

$$L_2 = \begin{bmatrix} 1.000000 & 0 & 0 & 0 \\ 0.176744 & 1.000000 & 0 & 0 \\ 0.129199 & 0.065359 & 1.000000 & 0 \\ 0.068399 & 0.034602 & 0.289975 & 1.0 \end{bmatrix}$$

$$R_2 = \begin{bmatrix} 8.6 & 4.222222 & 1.470588 & 1.000000 \\ 0 & 1.098191 & 0.034200 & 0.023256 \\ 0 & 0 & 4.422145 & 1.647059 \\ 0 & 0 & 0 & 2.394366 \end{bmatrix}$$

$$A_3 = R_2 L_2 = \begin{bmatrix} 9.604651 & 4.352941 & 1.760563 & 1.000000 \\ 0.200108 & 1.101231 & 0.040943 & 0.023256 \\ 0.683995 & 0.346021 & 4.899751 & 1.647059 \\ 0.163773 & 0.082850 & 0.694307 & 2.394366 \end{bmatrix} \tag{9.44}$$

The quadratic convergence method to be described in the next topic begins with matrix A_3.

A QUADRATIC CONVERGENT METHOD

The LR method is a good method for starting the process and should be used through enough iterations to guarantee that diagonal elements of the matrix A_k are ordered in absolute value and that the subdiagonal elements show a definite tendency to be converging to zero. At this point a different method is applied that has a quadratic convergence characteristic. The LR

method has produced the matrix:

$$
A_k = \begin{bmatrix}
a_{11} & a_{12} & a_{13} & \cdots & a_{1n} \\
a_{21} & a_{22} & a_{23} & \cdots & a_{2n} \\
 & & & \cdots & \\
\vdots & \vdots & \vdots & \cdots & \vdots \\
 & & & \cdots & \\
a_{n1} & a_{n2} & a_{n3} & & a_{nn}
\end{bmatrix}
\tag{9.45}
$$

From this matrix form a set of simultaneous linear equations by discarding column one as show in equation 9.46 and using it as the vector on the right side of equation 9.46

$$
\begin{bmatrix}
a_{22}-a_{11} & a_{23} & a_{24} & \cdots & a_{2n} \\
 & a_{33}-a_{11} & a_{34} & \cdots & a_{3n} \\
 & & a_{44}-a_{11} & \cdots & a_{4n} \\
 & & & & \vdots \\
 & & & & a_{nn}-a_{11}
\end{bmatrix}
\begin{bmatrix}
X_2 \\ X_3 \\ X_4 \\ \vdots \\ X_n
\end{bmatrix}
=
\begin{bmatrix}
-a_{21} \\ -a_{31} \\ -a_{41} \\ \vdots \\ -a_{n1}
\end{bmatrix}
\tag{9.46}
$$

Application of equation 9.46 to the matrix A_3 which resulted after two steps of the LR process gives equation 9.47 (see equation 9.44).

$$
\begin{bmatrix}
(1.101231-9.604651) & 0.040943 & 0.023256 \\
 & (4.899751-9.604651) & 1.647059 \\
 & & (2.394366-9.604651)
\end{bmatrix}
\begin{bmatrix}
X_2 \\ X_3 \\ X_4
\end{bmatrix}
$$

$$
=
\begin{bmatrix}
-0.200108 \\
-0.683995 \\
-0.163773
\end{bmatrix}
\tag{9.47}
$$

$$
\begin{bmatrix}
-8.503420 & 0.040943 & 0.023256 \\
 & -4.704900 & 1.647059 \\
 & & -7.210285
\end{bmatrix}
\begin{bmatrix}
X_2 \\ X_3 \\ X_4
\end{bmatrix}
=
\begin{bmatrix}
-0.200108 \\ -0.683995 \\ -0.163773
\end{bmatrix}
$$

$$(9.48)$$

Solving for the unknowns gives

$$X_2 = 0.024333$$

$$X_3 = 0.153331$$

$$X_4 = 0.022714$$

Define a transformation and its inverse by using X_2, X_3, and X_4.

$$
L_3 =
\begin{bmatrix}
1.000000 & 0 & 0 & 0 \\
0.024333 & 1.0 & 0 & 0 \\
0.153331 & 0 & 1.0 & 0 \\
0.022714 & 0 & 0 & 1.0
\end{bmatrix}
$$

$$(9.49)$$

$$
L_3^{-1} =
\begin{bmatrix}
1.000000 & 0 & 0 & 0 \\
-0.024333 & 1.0 & 0 & 0 \\
-0.153331 & 0 & 1.0 & 0 \\
-0.022714 & 0 & 0 & 1.0
\end{bmatrix}
$$

Using these transformations modify matrix A_3 as shown in equation 9.50; A_3 was obtained by two applications of the LR transformation process which precedes this topic.

$$A_4 = L_3^{-1} A_3 L_3$$

$$(9.50)$$

$$
A_4 =
\begin{bmatrix}
10.003233 & 4.352941 & 1.760563 & 1.000000 \\
-0.009698 & 0.995312 & -0.001896 & -0.001077 \\
-0.052698 & -0.321421 & 4.629802 & 1.493728 \\
+0.099419 & -0.016021 & 0.654318 & 2.371653
\end{bmatrix}
$$

$$(9.51)$$

The next transformation is formed by discarding the first two columns in matrix A_4 and forming the set of simultaneous equations of 9.52:

$$
\begin{bmatrix}
(a_{33}-a_{22}) & a_{34} & a_{35} & \cdots & a_{3n} \\
 & (a_{44}-a_{22}) & a_{45} & \cdots & a_{4n} \\
 & & (a_{55}-a_{22}) & \cdots & a_{5n} \\
 & & & \vdots & \\
 & & & & (a_{nn}-a_{22})
\end{bmatrix}
\begin{bmatrix} y_3 \\ y_4 \\ y_5 \\ \vdots \\ y_n \end{bmatrix}
=
\begin{bmatrix} -a_{32} \\ -a_{42} \\ -a_{52} \\ \vdots \\ -a_{n2} \end{bmatrix}
\tag{9.52}
$$

Substitution of the values from equation 9.51 into the general equation 9.52 gives

$$
\begin{bmatrix}
(4.629802-0.995312) & 1.493728 \\
 & (2.371653-0.995312)
\end{bmatrix}
\begin{bmatrix} y_3 \\ y_4 \end{bmatrix}
=
\begin{bmatrix} 0.321421 \\ 0.016021 \end{bmatrix}
\tag{9.53}
$$

$$y_3 = 0.083652$$

$$y_4 = 0.011640$$

Using these values to define the next transformation gives

$$
L_4 =
\begin{bmatrix}
1.0 & 0 & 0 & 0 \\
0 & 1.000000 & 0 & 0 \\
0 & 0.083652 & 1.0 & 0 \\
0 & 0.011640 & 0 & 1.0
\end{bmatrix}
$$

and

$$L_4^{-1} = \begin{bmatrix} 1.0 & 0 & 0 & 0 \\ 0 & 1.000000 & 0 & 0 \\ 0 & -0.083652 & 1.0 & 0 \\ 0 & -0.011640 & 0 & 1.0 \end{bmatrix}$$

$$A_5 = L_4^{-1} A_5 L_4$$

$$A_5 = \begin{bmatrix} 10.003233 & 4.511856 & 1.760563 & 1.000000 \\ -0.009698 & 0.995141 & -0.001896 & -0.001077 \\ -0.051887 & 0.000013 & 4.629961 & 1.493818 \\ 0.099532 & 0.054737 & 0.654340 & 2.371666 \end{bmatrix} \quad (9.54)$$

Note. It is the author's opinion that there is a typographical error in Ref. 6 in which the element a_{33} of A_5 is incorrectly given as 4.269961 on p. 61; it should actually be 4.629961.

Discarding the first three columns of the matrix A_5 gives the general plan

$$\begin{bmatrix} (a_{44} - a_{33}) & a_{45} & a_{46} & \cdots & a_{4n} \\ & (a_{55} - a_{33}) & a_{56} & \cdots & a_{5n} \\ & & (a_{66} - a_{33}) & \cdots & a_{6n} \\ & & & \vdots & \vdots \\ & & & & a_{nn} - a_{33} \end{bmatrix} \begin{bmatrix} z_4 \\ z_5 \\ z_6 \\ \vdots \\ z_n \end{bmatrix} = \begin{bmatrix} -a_{43} \\ -a_{53} \\ -a_{63} \\ \vdots \\ -a_{n3} \end{bmatrix}$$

$$(9.55)$$

Substitution of the values from (9.54) gives equation 9.56.

$$|(2.371666 - 4.629961)||z_4| = |-0.65430| \quad (9.56)$$

$$z_4 = 0.289750$$

The transformation is

$$L_5 = \begin{bmatrix} 1.0 & 0 & 0 & 0 \\ 0 & 1.0 & 0 & 0 \\ 0 & 0 & 1.000000 & 0 \\ 0 & 0 & 0.289750 & 1.0 \end{bmatrix}$$

and

$$L_5^{-1} = \begin{bmatrix} 1.0 & 0 & 0 & 0 \\ 0 & 1.0 & 0 & 0 \\ 0 & 0 & 1.000000 & 0 \\ 0 & 0 & -0.289750 & 1.0 \end{bmatrix}$$

$$A_6 = L_5^{-1} A_5 L_5 \tag{9.57}$$

$$A_6 = L_5^{-1} A_5 L_5 = \begin{bmatrix} 10.003233 & 4.511856 & 2.050313 & 1.000000 \\ -0.009698 & 0.995141 & -0.002208 & -0.001077 \\ -0.051887 & 0.000013 & 5.062795 & 1.493818 \\ 0.114566 & 0.054733 & -0.125414 & 1.938832 \end{bmatrix}$$

This completes the first "sweep." The second sweep returns to equation 9.46 and repeats the process.

$$L_6 = \begin{bmatrix} 1.000000 & 0 & 0 & 0 \\ -0.001077 & 1.0 & 0 & 0 \\ -0.006207 & 0 & 1.0 & 0 \\ -0.014206 & 0 & 0 & 1.0 \end{bmatrix}$$

$$L_7 = \begin{bmatrix} 1.0 & 0 & 0 & 0 \\ 0 & 1.000000 & 0 & 0 \\ 0 & -0.010601 & 1.0 & 0 \\ 0 & 0.010125 & 0 & 1.0 \end{bmatrix}$$

and

$$L_8 = \begin{bmatrix} 1.0 & 0 & 0 & 0 \\ 0 & 1.0 & 0 & 0 \\ 0 & 0 & 1.000000 & 0 \\ 0 & 0 & -0.049047 & 1.0 \end{bmatrix}$$

The three transformations complete the second sweep and the matrix 9.58 results.

$$A_9 = \begin{bmatrix} 9.999853 & 4.500246 & 2.001266 & 1.0 \\ -0.000002 & 1.000000 & 0 & 0 \\ -0.000022 & 0 & 5.001949 & 1.50025 \\ 0.000770 & 0.001638 & -0.003608 & 1.998198 \end{bmatrix} \quad (9.58)$$

After the third sweep the matrix is

$$A_{12} = \begin{bmatrix} 9.999997 & 4.499999 & 2.000001 & 1.000000 \\ -0.000002 & 1.000000 & 0 & 0 \\ 0.000002 & 0.000001 & 5.000003 & 1.500001 \\ 0.000001 & -0.000002 & -0.000002 & 2.000000 \end{bmatrix} \quad (9.59)$$

The latent roots are the diagonal elements 9.999997, 1.0, 5.000003, and 2.0.

In many engineering problems the eigenvalues are complex. By making a modification in the method that has just been described one can evaluate complex eigenvalues.

Consider the matrix of Section 5 of Ref. 6.

$$A_1 = \begin{bmatrix} 4 & -5 & 0 & 3 \\ 0 & 4 & -3 & -5 \\ 5 & -3 & 4 & 0 \\ 3 & 0 & 5 & 4 \end{bmatrix} \quad (9.60)$$

By the LR process of decomposing into L and R matrices and forming the product with R and L in the reverse order as has been described, the following is given.

$$A_2 = \begin{bmatrix} 6.25000 & -2.18750 & 3.64078 & 3.0000 \\ -7.50000 & -3.12500 & -9.06796 & -5.0000 \\ 8.28125 & 5.52344 & 6.81675 & 0.3125 \\ 4.54369 & 5.67961 & 7.35225 & 6.05825 \end{bmatrix} \tag{9.61}$$

$$A_3 = \begin{bmatrix} 15.88000 & -11.31304 & 4.39889 & 3.00000 \\ -0.34400 & 2.90261 & -5.05282 & -1.40000 \\ -10.63235 & 14.38518 & -6.33355 & -5.71304 \\ 2.58150 & -4.48956 & 0.89734 & 3.55094 \end{bmatrix} \tag{9.62}$$

$$A_4 = \begin{bmatrix} 13.66750 & -3.03182 & 2.86547 & 3.00000 \\ 3.04469 & -8.71587 & -4.27515 & -1.33501 \\ -6.28388 & 24.15839 & 9.46129 & -0.28312 \\ 0.25800 & -1.58285 & -0.81122 & 1.58707 \end{bmatrix} \tag{9.63}$$

$$A_5 = \begin{bmatrix} 11.73129 & -10.57540 & 2.80131 & 3.00000 \\ 0.43008 & 5.49061 & -4.87064 & -2.00332 \\ 1.35383 & 8.00050 & -3.03460 & -4.57567 \\ 0.03422 & 0.34395 & -0.03877 & 1.81271 \end{bmatrix} \tag{9.64}$$

Note. The element a_{11} of matrix A_5 on p. 63 of Ref. 6 also has a typographical error and is given as 11.70129 rather than 11.73129 as given on p. 57 of the same reference.

It is observed that the lower elements show no consistent tendency to become small, while the diagonal elements a_{11} and a_{44} show some stability, but the diagonal elements a_{22} and a_{33} are very unstable. It is concluded from elements a_{11} and a_{44} that there are real eigenvalues close to 11.7 and 1.8 and a conjugate pair of complex roots corresponding to the minor.

$$\begin{bmatrix} a_{22} & a_{23} \\ a_{32} & a_{33} \end{bmatrix} = \begin{bmatrix} 5.49 - 4.87 \\ 8.00 - 3.03 \end{bmatrix} \tag{9.65}$$

The characteristic equation of this minor is

$$\begin{bmatrix} 5.49 - \lambda & -4.87 \\ 8.00 & -3.03 - \lambda \end{bmatrix} \approx 22.33 - 2.46\lambda + \lambda^2$$

and the latent roots are in the neighborhood of $\lambda \approx 1.23 \pm 4.5i$.

Since several transformations of the *LR* type have been tried, the process of equations 9.50 is applied to columns that are suspected of having real eigenvalues, and a modification of the process is applied to two columns that evidently represent a conjugate pair of roots. Returning to matrix A_5 and applying equations 9.46–9.48, a transformation is defined

$$L_5 = \begin{bmatrix} 1.00000 & 0 & 0 & 0 \\ -0.00205 & 1.0 & 0 & 0 \\ 0.08951 & 0 & 1.0 & 0 \\ 0.00345 & 0 & 0 & 1.0 \end{bmatrix}$$

and

$$L_5^{-1} = \begin{bmatrix} 1.000000 & 0 & 0 & 0 \\ 0.00205 & 1.0 & 0 & 0 \\ -0.08951 & 0 & 1.0 & 0 \\ -0.00345 & 0 & 0 & 1.0 \end{bmatrix}$$

$$A_6 = L_5^{-1} A_5 L_5 = \begin{bmatrix} 12.01406 & -10.57540 & 2.80131 & 3.00000 \\ 0.00057 & 5.46893 & -4.86490 & -1.99717 \\ -0.02537 & 8.94710 & -3.28535 & -4.84420 \\ -0.00515 & 0.38044 & -0.04843 & 1.80236 \end{bmatrix}$$

$$\tag{9.66}$$

The process to be used on the two columns suspected of having a conjugate pair of roots is as follows. Assume that columns 2 and 3 are to be modified; form the set of equations 9.67 and solve for the unknowns.

$$
\begin{bmatrix}
a_{44}-a_{22} & -a_{32} & a_{45} & 0 & a_{46} & 0 & \cdots \\
-a_{23} & a_{44}-a_{33} & 0 & a_{45} & 0 & a_{46} & \cdots \\
 & & a_{55}-a_{22} & -a_{32} & a_{56} & 0 & \cdots \\
 & & -a_{23} & a_{55}-a_{33} & 0 & a_{56} & \\
 & & & & & & \vdots
\end{bmatrix}
\begin{bmatrix} X_4 \\ y_4 \\ X_5 \\ y_5 \\ \vdots \end{bmatrix}
=
\begin{bmatrix} -a_{42} \\ -a_{43} \\ -a_{52} \\ -a_{53} \\ \vdots \end{bmatrix}
$$

(9.67)

For the example, take the values from A_6 to give equation 9.68

$$
\begin{bmatrix}
(1.80236-5.46893) & -8.94710 \\
+4.86490 & (1.80236+3.28535)
\end{bmatrix}
\begin{bmatrix} X_4 \\ y_4 \end{bmatrix}
=
\begin{bmatrix} -0.38044 \\ +0.04843 \end{bmatrix}
$$

(9.68)

$$
\begin{bmatrix}
-3.66657 & -8.94710 \\
+4.86490 & 5.08771
\end{bmatrix}
\begin{bmatrix} X_4 \\ y_4 \end{bmatrix}
=
\begin{bmatrix} -0.38044 \\ 0.04843 \end{bmatrix}
$$

$$X_4 = -0.06040$$

$$y_4 = 0.06727$$

$$
L_6 =
\begin{bmatrix}
1.0 & 0 & 0 & 0 \\
0 & 1.000000 & 0 & 0 \\
0 & 0 & 1.00000 & 0 \\
0 & -0.06040 & 0.06727 & 1.0
\end{bmatrix}
$$

$$
A_7 = L_6^{-1} A_6 L_6 =
\begin{bmatrix}
12.01406 & -10.75660 & 3.00312 & 3.00000 \\
0.00057 & 5.58956 & -4.99925 & -1.99717 \\
-0.02537 & 9.23969 & -3.61122 & -4.84420 \\
-0.00341 & -0.01237 & 0.01379 & 2.00760
\end{bmatrix}
$$

(9.69)

One more sweep gives the matrix A_9.

$$A_9 = \begin{bmatrix} 12.00005 & -10.76197 & 3.00000 & 3.00000 \\ 0.00002 & 5.60325 & -4.99999 & -1.99999 \\ -0.00005 & 9.23803 & -3.60331 & -4.84132 \\ -0.00003 & 0 & 0.00002 & 2.00002 \end{bmatrix}$$

The real roots are 12.0 and 2.0 with a conjugate pair determined from evaluation of the minor

$$\det \begin{bmatrix} 5.60325 - \lambda & -4.99999 \\ 9.23803 & -3.60331 - \lambda \end{bmatrix} = 0$$

$$\lambda = 1 \pm 5i$$

APPLICATIONS

The applications of eigenvalues are many and varied, from the flutter of an airplane wing [2] to oscillations in a power system produced by a pulsating customer load [8], the interaction between the mechanical and electrical system [9], or the optimization of adjustments of a control system [10]. The applications will surely increase, and it is essential that an engineer has at least an elementary understanding of how the roots are determined after the system being considered has been described by the matrix A.

LINEAR PROGRAMMING

Linear programming deals with a special type of problem in which a maximum (or a minimum) of a certain linear function is required, subject to certain inequality constraints between the variables. Engineers generally have a reputation for solving large sets of rather complicated equations, but little was done to solve sets of inequalities until relatively recently. In 1947, the first work was reported by Dantzig when he described the simplex method of solution of minimizing a cost function (objective function) subject to equality and inequality constraints [11]. Since that time, algorithms have been developed that will handle problems with thousands of conditions. See the references listed in the book by Beale [12]. If the reader encounters a large problem of this type, he should use one of the very sophisticated linear programming routines that are available at any of the many service bureaus. This section of the book explains the details of how the simplex method achieves the minimization of an objective function in enough detail that a small linear programming code could be written.

THE CONSTRAINTS

The equality and inequality constraints define a region in the multidimensional space. Any point in the region or on the boundary satisfies all of the conditions that have been imposed. In maximizing an objective function one must find the point in space that defines the vector that produces the maximum value of the objective function. If a minimum is sought, it is the vector that produces the smallest value of the required objective function.

Since visualizing a three-dimensional region is difficult and a higher order space is impossible, the illustrations are limited to two dimensions. A maximum value of the objective function always occur at a vertex of the space, the point where two constraints intersect. In small problems it is possible to test the objective function for each set of points at all the verticies of the polyhedron defining the space, but the number of verticies increases much more rapidly than the number of constraints; therefore this method is not satisfactory for problems with many constraints.

The simplex method starts with some vertex of the polyhedron, and at each step proceeds to the neighboring vertex which increases the value of the objective function the greatest amount. The final solution has been found when no neighboring vertex gives a greater value to the objective function.

Example 1 See Fig. 9.3. The objective function to be maximized is $C = 3X_1 + 2X_2$

Constraints are as follows:

$$X_1 - X_2 \leqslant 2$$

$$30X_1 - 10X_2 \leqslant 72$$

$$5X_1 + X_2 \leqslant 20 \qquad (9.70)$$

$$4X_1 + 3X_2 \leqslant 23$$

$$-X_1 + 2X_2 \leqslant 4$$

A "slack variable" is added to each inequality to convert the constraints to equalities. The slack variables are not permitted to take on negative values.

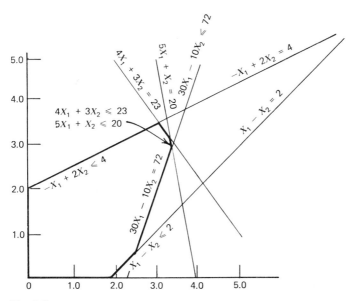

Fig. 9.3.

$$C = 3X_1 + 2X_2$$

$$X_1 - X_2 + X_3 \qquad\qquad\qquad = 2$$

$$30X_1 - 10X_2 \;\; + X_4 \qquad\qquad\qquad = 72$$

$$5X_1 + X_2 \qquad\qquad\quad + X_5 \qquad\qquad = 20 \qquad\qquad (9.71)$$

$$4X_1 + 3X_2 \qquad\qquad\qquad\quad + X_6 \qquad\quad = 23$$

$$- X_1 + 2X_2 \qquad\qquad\qquad\qquad\quad + X_7 \;\; = 4$$

Solve this set of equations for the objective function and the slack variables in terms of the actual variables.

$$C = 3X_1 + 2X_2$$

$$X_3 = 2 - X_1 + X_2$$

$$X_4 = 72 - 30X_1 + 10X_2 \tag{9.72}$$

$$X_5 = 20 - 5X_1 - X_2$$

$$X_6 = 23 - 4X_1 - 3X_2$$

$$X_7 = 4 + X_1 - 2X_2$$

To begin the process, a "feasible solution" is required. In this problem, $X_1 = X_2 = 0$ is a feasible solution, since the basic variables (those on the left of equation 9.72) are all positive as required. The search begins at the vertex that is the origin of the coordinate system. It can be seen that C will increase if either X_1 or X_2 increases, and furthermore that it will increase faster for changes in X_1 than for X_2, since the coefficient of X_1 is 3 compared to 2 for X_2 in the objective function. Therefore, X_1 should enter the basis (be permitted to take on a nonzero value). As X_1 is allowed to increase, X_3 will be the first basic variable to become zero. Then X_3 should leave the basis and be set equal to zero: $X_1 = 2 + X_2 - X_3$. The set of equations then can be written by substituting this value for X_1 in all the other equations.

$$C = 3X_1 + 2X_2 = 3(2 + X_2 - X_3) + 2X_2 = 6 + 5X_2 - 3X_3$$

$$X_1 = 2 + X_2 - X_3$$

$$X_4 = 72 - 30(2 + X_2 - X_3) + 10X_2 = 12 - 20X_2 + 30X_3 \tag{9.73}$$

$$X_5 = 20 - 5(2 + X_2 - X_3) - X_2 = 10 - 6X_2 + 5X_3$$

$$X_6 = 23 - 4(2 + X_2 - X_3) - 3X_2 = 15 - 7X_2 + 4X_3$$

$$X_7 = 4 + (2 + X_2 - X_3) - 2X_2 = 6 - X_2 - X_3$$

The objective function has increased from zero to 6.0, but it can be seen that it would be further increased if X_2 takes on a positive (nonzero) value. The maximum permissable value for X_2 is 0.6, because at that value X_4 becomes zero and any further increase in X_2 would cause X_4 to be negative which is counter to the restriction on slack variables. Solving for X_2 gives equation 9.74 in which X_2 enters and X_4 leaves the basis.

$$20X_2 = 12 + 30X_3 - X_4 \tag{9.74}$$

or

$$X_2 = 0.6 + 1.5X_3 - 0.05X_4$$

Substitution of this expression into the other equations gives

$$C = 6 + 5(0.6 + 1.5X_3 - 0.05X_4) - 3X_3 = 9 + 4.5X_3 - 0.25X_4$$

$$X_1 = 2 + (0.6 + 1.5X_3 - 0.05X_4) - X_3 = 2.6 + 0.5X_3 - 0.05X_4$$

$$X_2 = 0.6 + 1.5X_3 - 0.05X_4 \tag{9.75}$$

$$X_5 = 10 - 6(0.6 + 1.5X_3 - 0.05X_4) + 5X_3 = 6.4 - 4X_3 + 0.3X_4$$

$$X_6 = 15 - 7(0.6 + 1.5X_3 - 0.05X_4) + 4X_3 = 10.8 - 6.5X_3 + 0.35X_4$$

$$X_7 = 6 - (0.6 + 1.5X_3 - 0.05X_4) - X_3 = 5.4 - 2.5X_3 + 0.05X_4$$

The objective function has increased to 9.0 and would be still further increased if X_3 would reenter the basis and would be limited to 1.6 at which value X_5 would be zero. The variable X_3 reenters the basis and X_5 becomes nonbasic.

$$4X_3 = 6.4 + 0.3X_4 - X_5$$

or

$$X_3 = 1.6 + 0.075X_4 - 0.25X_5$$

Substitution of this value for X_3 in the equations gives

$$C = 16.2 + 0.0875X_4 - 1.125X_5$$

$$X_1 = 3.4 - 0.0125X_4 - 0.125X_5$$

$$X_2 = 3.0 + 0.0625X_4 - 0.375X_5 \tag{9.76}$$

$$X_3 = 1.6 + 0.075X_4 - 0.25X_5$$

$$X_6 = 0.4 - 0.1375X_4 + 1.625X_5$$

$$X_7 = 1.4 - 0.1375X_4 + 0.625X_5$$

A further increase in the objective function is possible if X_4 returns to the basis. In this case X_6 becomes zero as the limiting condition when $X_4 = 2.90909090$. Solving for X_4 gives

$$0.1375X_4 = 0.4 + 1.625X_5 - X_6$$

$$X_4 = 2.909090 + 11.81818X_5 - 7.272727X_6$$

Substitution of this value in equation 9.73 gives

$$C = 16.4545 - 0.909090X_5 - 0.636363X_6$$

$$X_1 = 3.363636 - 0.272727X_5 + 0.090909X_6$$

$$X_2 = 3.181818 + 0.363636X_5 - 0.454545X_6 \qquad (9.77)$$

$$X_3 = 1.81818 + 0.636363X_5 - 0.545454X_6$$

$$X_4 = 2.90909 + 11.81818X_5 - 7.272727X_6$$

$$X_7 = 1.0 - 1.0X_5 + 1.0X_6$$

No further improvement is possible, since the coefficients of both X_5 and X_6 are negative in the objective function and increasing either would decrease C. The objective function is equal to 16.4545 when $X_1 = 3.363636$ and $X_2 = 3.181818$. The process began at $X_1 = X_2 = 0$ and progressed across the X_1 axis to the point $X_1 = 2.0$ and $X_2 = 0$. Then up the line $X_1 - X_2 = 2$ to the point $X_1 = 2.6$ and $X_2 = 0.6$. The next iteration followed the line $30X_1 - 10X_2 = 72$ to the point $X_1 = 3.4$ and $X_2 = 3.0$. The final step followed the line $5X_1 + X_2 = 20$ to $X_1 = 3.3636$ and $X_2 = 3.1818$ (see Fig. 9.3).

It seems that the process has started at one vertex and progressed systematically to the next vertex and the next. In two dimensions, this is exactly what happens, but in a multidimensional space, the process skips to the vertex of all the neighboring vertices that gives the greatest improvement. Since there are a finite number of vertices, the solution will be obtained in a finite number of steps. "Degeneracy" can occur if one or more basic variables equal zero at some step in the solution; an infinite looping can occur in the problem when the same vertex is represented by different expressions. When it does occur, the problem can be restated by making very minor changes in some of the constants in the equations that are causing the difficulty.

Example 1 was very simple since a feasible solution was obtained immediately by equating the genuine variables (X_1 and X_2) to zero; then all slack variables were nonnegative as required. However, such a simple case rarely occurs and a feasible solution must be obtained before the process can begin.

Example 2 See Fig. 9.4.

$$C = 2X_1 + 3X_2$$
$$-X_1 + 2X_2 \leqslant 6$$
$$9X_1 + 8X_2 \leqslant 72$$
$$3X_1 - 4X_2 \leqslant 12 \tag{9.78}$$
$$X_1 + 3X_2 \geqslant 6$$
$$-3X_1 + 2X_2 \leqslant 2$$
$$2X_1 - X_2 = 10$$

Slack variables are introduced to convert the constraints to equalities. Solve the equations in terms of the slack variables and define a new variable $X_0 = -C$.

$$X_0 = -2X_1 - 3X_2 \qquad \text{(objective function)}$$
$$X_3 = 6 + X_1 - 2X_2$$
$$X_4 = 72 - 9X_1 - 8X_2$$
$$X_5 = 12 - 3X_1 + 4X_2 \tag{9.79}$$
$$X_6 = -6 + X_1 + 3X_2$$
$$X_7 = 2 + 3X_1 - 2X_2$$
$$10 = 2X_1 - X_2$$

A feasible solution is no longer obtained by setting the genuine variables to zero, because the equation $X_6 = -6 + X_1 + 3X_2$ will violate the restriction that slack variables cannot be negative. Furthermore the last equation, $2X_1 - X_2 = 10$, is not satisfied by $X_1 = X_2 = 0$.

If the restriction that slack variables can not have negative values is removed, then the only difficulty remains with the last equation, when $X_1 = X_2 = 0$. Define an artificial variable A_1 which is equal to the difference between the left- and right-hand sides of the last equation. The sign is chosen so that A_1 is positive. When for $X_1 = X_2 = 0$, the equations to be

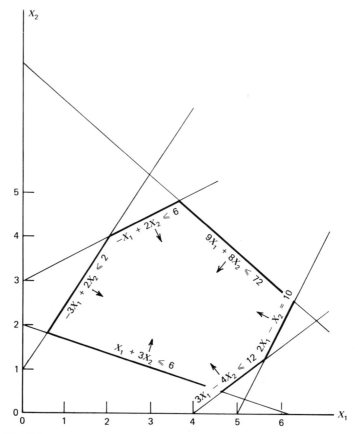

Fig. 9.4.

considered are as follows.

$$X_0 = -2X_1 - 3X_2$$
$$X_3 = 6 + X_1 - 2X_2$$
$$X_4 = 72 - 9X_1 - 8X_2$$
$$X_5 = 12 - 3X_1 + 4X_2$$
$$X_6 = -6 + X_1 + 3X_2 \qquad (9.80)$$
$$X_7 = 2 + 3X_1 - 2X_2$$
$$A_1 = 10 - 2X_1 + X_2$$

In 1965 Wolfe [13] devised a modification of the simplex method which permits slack variable to have negative values, and only equality constraints required artificial variables. His new rules for optimizing the objective function are to continue increasing the selected nonbasic variable until one of the following happens.

1. An artificlal variable becomes zero.
2. A basic variable that is positive becomes zero.
3. The selected variable could be increased indefinitely without forcing any variable to zero.

Select X_2 as the variable to enter the basis, since it will change X_0 faster than X_1. The limitation is X_7, since it becomes zero quicker than any other variable (rule 2). Exchange X_2 and X_7 in the basis.

$$2X_2 = 2 + 3X_1 - X_7$$

$$X_2 = 1 + 1.5X_1 - 0.5X_7$$

Substitution of this value in the equations gives

$$X_0 = -2X_1 - 3(1 + 1.5X_1 - 0.5X_7) = -3 - 6.5X_1 + 1.5X_7$$

$$X_3 = 6 + X_1 - 2(1 + 1.5X_1 - 0.5X_7) = 4 - 2X_1 + 1.0X_7$$

$$X_4 = 72 - 9X_1 - 8(1 + 1.5X_1 - 0.5X_7) = 64 - 21X_1 + 4X_7$$

$$X_5 = 12 - 3X_1 + 4(1 + 1.5X_1 - 0.5X_7) = 16 + 3X_1 - 2X_7 \qquad (9.81)$$

$$X_6 = -6 + X_1 + 3(1 + 1.5X_1 - 0.5X_7) = -3 + 5.5X_1 - 1.5X_7$$

$$X_2 = 1 + 1.5X_1 - 0.5X_7$$

$$A_1 = 10 - 2X_1 + 1 + 1.5X_1 - 0.5X_7 = 11 - 0.5X_1 - 0.5X_7$$

Now X_1 should be increased and X_3 will be the limiting variable. Therefore, X_1 should enter and X_3 should leave the basis.

$$2X_1 = 4 - X_3 + X_7$$

$$X_1 = 2 - 0.5X_3 + 0.5X_7$$

Substitution of this value in the equations gives

$$X_0 = -3 - 6.5(2 - 0.5X_3 + 0.5X_7) + 1.5X_7 = -16 + 3.25X_3 - 1.75X_7$$

$$X_1 = 2 - 0.5X_3 + 0.5X_7$$

$$X_4 = 64 - 21(2 - 0.5X_3 + 0.5X_7) + 4X_7 = 22 + 10.5X_3 - 6.5X_7$$

$$X_5 = 16 + 3(2 - 0.5X_3 + 0.5X_7) - 2X_7 = 22 - 1.5X_3 - 0.5X_7$$

$$X_6 = -3 + 5.5(2 - 0.5X_3 + 0.5X_7) - 1.5X_7 = 8 - 2.75X_3 + 1.25X_7$$

$$X_2 = 1 + 1.5(2 - 0.5X_3 + 0.5X_7) - 0.5X_7 = 4 - 0.75X_3 + 0.25X_7$$

$$A_1 = 11 - 0.5(2 - 0.5X_3 + 0.5X_7) - 0.5X_7 = 10 + 0.25X_3 - 0.75X_7$$

Now X_0 can be decreased by increasing X_7 limited by X_4, since it must not become negative.

$$+6.5X_7 = 22 - X_4 + 10.5X_3$$

$$X_7 = 3.38462 - 0.15385X_4 + 1.61538X_3$$

Substitution of this value in the equations gives

$$X_0 = -16 + 3.25X_3 - 1.75(3.38462 - 0.15385X_4 + 1.61538X_3)$$

$$= -21.92 + 0.423085X_3 + 0.26924X_4$$

No further improvement is possible, since increasing either X_3 or X_4 would increase X_0. It is only necessary to determine the value of the genuine variables.

$$X_1 = 2.0 - 0.5X_3 + 0.5X_7 = 2.0 - 0.5X_3 + 0.5(3.38462 - 0.15385X_4 + 1.61538X_3)$$

$$X_1 = 3.69231 - 0.076925X_4 + 0.30769X_3 = 3.69231 \qquad (9.83)$$

$$X_2 = 4 - 0.75X_3 + 0.25(3.38462 - 0.15385X_4 + 1.61538X_3)$$

$$= 4.84616 - 0.0384625X_4 - 1.15385X_3 = 4.84616$$

At the end of the problem the objective function and the genuine variables are given by equation 9.84:

$$X_0 = -21.92000$$

$$X_1 = \quad 3.69231 \tag{9.84}$$

$$X_2 = \quad 4.84616$$

(see Fig. 9.4).

It is of interest to look at the simplex method as a series of matrix operations to see the relationship between the Shipley inversion method and the simplex algorithm. Setting up the matrix of the coefficients of equations 9.80 gives the matrix 9.85

$$
\begin{array}{c}
 & 1 & X_1 & X_2 \\
\begin{array}{c}
X_0 \\
X_3 \\
X_4 \\
X_5 \\
X_6 \\
X_7 \\
A_1
\end{array}
&
\left[
\begin{array}{ccc}
0 & -2 & -3 \\
6 & 1 & -2 \\
72 & -9 & -8 \\
12 & -3 & 4 \\
-6 & 1 & 3 \\
2 & 3 & -2 \\
10 & -2 & 1
\end{array}
\right]
\end{array}
\tag{9.85}
$$

The element with the most negative coefficient in row X_0 is in column X_2 which is therefore chosen as the pivoting column (X_2 will enter the basis, since $a_{02} = -3$). To find the variable to leave the basis, consider only negative values in the X_2 column and obtain the minimum ratio $a_{n0}/-a_{n2}$, $n \neq 0$. It is found that $a_{70}/-a_{72} = 2/2 = 1.0$ is the smallest; therefore, X_7 will leave the basis and X_2 will enter the basis.

In the pivot column (column $q = X_2$) define

$$a'_{iq} = \frac{a_{iq}}{a_{pq}}$$

The $i \neq p$ and p is the pivot row. This differs from the inversion algorithm by a minus sign (see equation 2.4).

In the pivot row, define (row $p = X_7$):

$$a'_{pj} = \frac{a_{pj}}{-a_{pq}} = \frac{a_{7j}}{-a_{72}} \qquad p \neq j$$

This equation agrees with equation 2.3.

All elements not in the pivot row and column are replaced by $a'_{ij} = a_{ij} - a_{iq} a_{pj} / a_{pq}$. This agrees with equation (2.2). The pivot element is replaced by its reciprocal, this differs from equation (2.5) by a minus sign. Applying these formulae to the matrix (9.85) gives matrix (9.86) but note the change in the heading of column 2 and row 6.

$$
\begin{array}{c}
 \\
X_0 \\
X_3 \\
X_4 \\
X_5 \\
X_6 \\
X_2 \\
A_1
\end{array}
\begin{array}{ccc}
1 & X_1 & X_7 \\
\left[\begin{array}{ccc}
-3 & -6.5 & 1.5 \\
4 & -2.0 & 1.0 \\
64 & -21.0 & 4.0 \\
16 & 3.0 & -2.0 \\
-3 & 5.5 & -1.5 \\
+1 & +1.5 & -0.5 \\
11 & -0.5 & -0.5
\end{array}\right]
\end{array}
\qquad (9.86)
$$

It can be seen that the X_1 column contains the next pivot because $a_{01} = -6.5$. Now the minimum ratio occurs in the X_3 row because the result of $a_{10} / -a_{11} = 4/2$, is smaller than $64/21$ and all other elements in the column are positive. The next pivoting operation is left to the reader. It must check with equations 9.82.

It is rather interesting that a technique (Shipley inversion) developed in an entirely different context is the same technique, except for a minus sign here and there, that is used in linear programming. In the inversion algorithm, the pivot is always a diagonal element, whereas in linear programming the pivot is determined by consideration of the matrix elements.

OPTIMIZATION

The optimization of a function to achieve an objective, for example, minimum cost of generation, has occupied the time and talent of many men in the power industry over a long period of time. In 1943 Steinberg and Smith [15] showed the necessity of all generators of a system being operated at the same incremental cost of power production. This fact can be readily seen by examination of Fig. 9.5 in which all generators of the system are operated at an incremental cost of λ, with the exception of one machine, which is being operated at an incremental cost of $\lambda + \Delta\lambda$. The incremental cost of generation is the increase of cost of production that will result from a small incremental change in generator load.

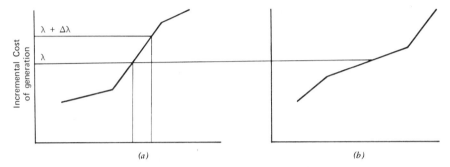

Fig. 9.5. (*a*) Single machine *MW*. (*b*) Remainder of system *MW*.

If the load on the single generator that is operating at an incremental cost of $\lambda + \Delta\lambda$ drops 1 MW of load it would do so at its incremental cost of $\lambda + \Delta\lambda$. This 1 MW of load would then be picked up by the other generators of the system at the incremental cost of λ.

The total cost of generation after this shift of one unit of generation would therefore be reduced by $\Delta\lambda$. The increase of generation furnished by the system would slightly increase the system λ and would decrease the cost discrepancy $\Delta\lambda$ of the single machine that is operating at an incremental rate of production above that of the system. This process could be repeated, and with each adjustment a savings would occur. Therefore, all generators should be forced to operate at the same incremental cost of power production λ.

E. E. George demonstrated the fact that the incremental costs of generation should be adjusted for transmission losses [16]. If the load of a generator is increased by ΔMW at a cost of $\lambda = dF/dP$ (rate of change of cost of fuel consumption), and if the system losses are increased as a result of the shift in generation, the increase in loses $\Delta L = dL/dP_1$ must be paid for at the incremental cost of production λ. Therefore, the cost of production at each plant must be increased by the cost of producing the incremental change in losses caused by increasing the power delivered to the system by the particular generator.

$$\frac{dF}{dP} + \lambda \frac{dL}{dP} = \lambda$$

Here λ is the Lagrangian multiplier, the use of which in optimization problems are discussed later in this chapter [17].

THE OPTIMIZATION PROBLEM

Again the illustrations are restricted to equations in two unknowns for easier understanding of the material although thousands of variables may be involved in a typical engineering problem. The determination of the minimum (or maximum) of a function by examining the location of values for which the derivative of the function vanishes is well known from the calculus. Great care must be used to distinguish between maxima, minima, and saddle points, that may be determined by consideration of only the first derivative of the function. Complete volumns [18–20] have been written on the subject. All that is attempted here is to briefly acquaint the reader with some of the methods being used. For applications, limitations, and pitfalls of the methods, the reader is referred to the abundant source material [22–29].

LAGRANGIAN METHOD

Suppose that one wants to minimize the function $f(X_1, X_2) = k$, subject to the constraint that $g(X_1, X_2) = 0$. That is, $f(X_1, X_2)$ is allowed to increase until it just touches the curve of $g(X_1, X_2) = 0$, and at this point the slope of f and g will be equal. Here it is assumed that the matter of convex and concave functions are not involved. That means that dX_1/dX_2 of $f(X_1, X_2)$ will be equal to dX_1/dX_2 of $g(X_1, X_2)$ at the point where the two curves touch. But

$$\frac{dX_1}{dX_2} = -\frac{\partial f/\partial X_2}{\partial f/\partial X_1} \quad \text{and} \quad \frac{dX_1}{dX_2} = -\frac{\partial g/\partial X_2}{\partial g/dX_1} \tag{9.87}$$

Therefore

$$\frac{\partial f/dX_2}{\partial f/\partial X_1} = \frac{\partial g/\partial X_2}{\partial g/\partial X_1} \tag{9.88}$$

which may be written

$$\frac{\partial f/\partial X_1}{\partial g/\partial X_1} = \frac{\partial f/\partial X_2}{\partial g/\partial X_2} = \lambda \tag{9.89}$$

This common ratio is called λ, the Lagrangian multiplier. Then

and

$$\left.\begin{array}{c} \dfrac{\partial f}{\partial X_1} - \lambda \dfrac{\partial g}{\partial X_1} = 0 \\[3mm] \dfrac{\partial f}{\partial X_2} - \lambda \dfrac{\partial g}{\partial X_2} = 0 \end{array}\right\} \qquad (9.90)$$

But these are exactly the same conditions that would be obtained if a new unconstrained function $h(X_1,X_2,\lambda) = f(X_1,X_2) - \lambda g(X_1,X_2)$ is minimized.

Notice that h is a function of three variables, and the conditions that must be satisfied for a minimum are

$$\left.\begin{array}{c} \dfrac{\partial h}{\partial X_1} = 0 \\[3mm] \dfrac{\partial h}{\partial X_2} = 0 \\[3mm] \dfrac{\partial h}{\partial \lambda} = 0 \end{array}\right\} \qquad (9.91)$$

These three simultaneous equations must be solved for the values X_1, X_2, and λ.

Example 1 Minimize the function $f(X_1,X_2) = X_1 + X_1 X_2 + 3X_2 = k$, subject to the constraint $g(X_1,X_2) = X_1^2 + X_2 = 8$ for positive values of the variables X_1, X_2. To minimize the constrained function, define an unconstrained function $h(X_1,X_2) = f(X_1,X_2) - \lambda g(X_1,X_2)$

$$h(X_1,X_2) = X_1 + X_1 X_2 + 3X_2 - \lambda(X_1^2 + X_2 - 8)$$

The extrema (maximum or minimum) will occur when all of the partial derivatives of h with respect to the three variables are equal to zero.

$$\frac{\partial h}{\partial X_1} = 1 + X_2 - 2\lambda X_1 = 0$$

$$\frac{\partial h}{\partial X_2} = X_1 + 3 - \lambda = 0$$

$$\frac{\partial h}{\partial \lambda} = -X_1^2 - X_2 + 8 = 0$$

Solution of this set of equations gives

$$X_1 = 1, \qquad X_2 = 7, \qquad \text{and } \lambda = 4$$

Thus

$$f(X_1,X_2)=X_1+X_1X_2+3X_2=k=29$$

The minimum value for $f(X_1,X_2)$ in the first quadrant subject to the constraint $g(X_1,X_2)$ is 29. By referring to Fig. 9.6 one can see that this value of k represents the member of the family of curves $f(X_1,X_2)$ that is tangent to the curve $g(X_1,X_2)$ at the point $X_1=1$, $X_2=7$. It is immediately seen that since one curve is convex and the other is concave the method is misleading, because it gives a relative minimum. The absolute minimum occurs when $X_1=2\sqrt{2}$ and $X_2=0$. For this pair of values $k=2\sqrt{2}$. This simple example illustrates two things—how the method is used and how the wrong answer of very simple problems will be obtained if care is not exercised in application of the method.

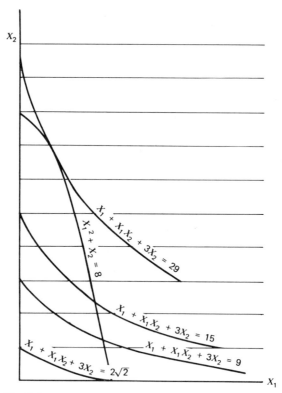

Fig. 9.6.

MULTIPLE CONSTRAINTS

If it is desired to obtain the maximum of a function of n unknowns $f(X_1, X_2, \ldots X_n)$ subject to m equality constraints $g_1(X_1, X_2, \ldots X_n)$, $g_2(X_1, X_2, \ldots X_n) \ldots g_m(X_1, X_2, \ldots X_n)$, a similar process is carried out. Form the unconstrained function

$$h(X_1, X_2, \ldots X_n, \lambda_1, \lambda_2, \ldots \lambda_m) = f(X_1, X_2, \ldots X_n)$$

$$+ \lambda_1 g_1(X_1, X_2, \ldots X_n) + \ldots + \lambda_m g_m(X_1, X_2, \ldots X_n) \quad (9.92)$$

Obtain the $n + m$ partial derivatives with respect to X_i, $i = 1$ to n, and λ_j, $j = 1$ to m. Solve this set of equations for the unknowns. Substitution of the values that have been determined into $f(X_1, X_2, \ldots X_n)$ gives the optimum value of the function subject to the constraints.

INEQUALITY CONSTRAINTS

To minimize a function $f(X_1, X_2)$ subject to the inequality constraint $g(X_1, X_2) \geqslant 0$, a similar method is used. A nonnegative term Z^2 is added to the inequality constraint $g(X_1, X_2) \geqslant 0$ which will only take on a value other than zero if the constraint $g(X_1, X_2) \geqslant 0$ is violated. That is, if $g(X_1, X_2) < 0$, then Z^2 will have the value required to satisfy equation 9.93;

$$g(X_1, X_2) + Z^2 = 0 \qquad (9.93)$$

If $g(X_1, X_2) \geqslant 0$, then $Z^2 = 0$.

Form the unconstrained function

$$h(X_1, X_2) = f(X_1, X_2) - \mu[g(X_1, X_2) + Z^2] = 0 \qquad (9.94)$$

The function h now has four variables, X_1, X_2, μ, and Z^2. Obtain the partial derivatives of h with respect to these variables and equate them to zero.

$$\frac{\partial h}{\partial X_1} = \frac{\partial f}{\partial X_1} - \mu \frac{\partial g}{\partial X_1} = 0$$

$$\frac{\partial h}{\partial X_2} = \frac{\partial f}{\partial X_2} - \mu \frac{\partial g}{\partial X_2} = 0$$

$$\frac{\partial h}{\partial \mu} = -g(X_1, X_2) - Z^2 = 0 \qquad (9.95)$$

$$\frac{\partial h}{\partial Z} = -2\mu Z = 0$$

The last condition means that either μ or Z, or both μ and Z must be equal to zero. This technique for treatment of inequality constraints was first described by Kuhn and Tucker [21].

Example 2 Minimize

$$f(X_1, X_2) = 2X_1 + 2X_1 X_2 + 3X_2 = k$$

subject to the inequality constraint

$$g(X_1, X_2) = X_1^2 + X_2 - 3 \geqslant 0$$

Form the unconstrained function

$$h = f(X_1, X_2) - \mu[g(X_1, X_2) + Z^2] = 2X_1 + 2X_1 X_2 + 3X_2$$
$$- \mu(X_1^2 + X_2 - 3 + Z^2) = k$$

The term Z^2 will be adjusted to guarantee that the right-hand term will be equal to zero.

$$\frac{\partial h}{\partial X_1} = 2 + 2X_2 - 2\mu X_1 = 0$$

$$\frac{\partial h}{\partial X_2} = 2X_1 + 3 - \mu = 0$$

$$\frac{\partial h}{\partial \mu} = X_1^2 + X_2 - 3 + Z^2 = 0$$

$$\frac{\partial h}{\partial Z} = 2\mu Z = 0$$

The last term requires that either μ or Z or both μ and Z must be equal to zero. A trial solution can be attempted by assuming $Z = 0$. Then

$$2 + 2X_2 - 2\mu X_1 = 0 \tag{9.96}$$

$$3 + 2X_1 - \mu = 0 \tag{9.97}$$

$$-3 + X_2 + X_1^2 = 0 \tag{9.98}$$

From equation 9.97

$$\mu = 3 + 2X_1$$

Substitution of this value into the equation 9.96 gives

$$2+2X_2-2X_1(3+2X_1)=2+2X_2-6X_1-4X_1^2=0$$

From the equation 9.98

$$X_2=3-X_1^2$$

Then

$$2+6-2X_1^2-6X_1-4X_1^2=0$$

$$6X_1^2+6X_1-8=0$$

$$X_1=\frac{-6\pm\sqrt{36+192}}{12}=\frac{-6\pm15.1}{12}=\frac{9.1}{12}=0.76$$

$$X_2=3-X_1^2=3-0.76^2=3-0.578=2.422$$

It can be seen from Fig. 9.7 that again we have a relative minimum, since $X_1=\sqrt{3}$ and $X_2=0$ give a smaller value of $f(X_1,X_2)$; $f(\sqrt{3},0)=\sqrt{3}$ while $f(0.76,2.422)=12.46$.

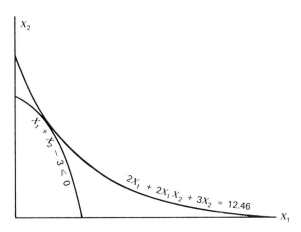

Fig. 9.7.

THE GRADIENT METHOD OF OPTIMIZATION

This method is also known as the method of steepest ascent. Assume it is desired to optimize a function

$$f(X_1,X_2,X_3,X_4)=X_1^3-2X_1+X_2^2+X_2-4X_3^4+6X_3^2-X_4^{-2}+5X_4=k$$

That is, the values of the four variables must be found which will result in a maximum value of k. The gradient

$$\nabla f = \begin{bmatrix} \dfrac{\partial f}{\partial X_1} \\ \dfrac{\partial f}{\partial X_2} \\ \dfrac{\partial f}{\partial X_3} \\ \dfrac{\partial f}{\partial X_4} \end{bmatrix} = \begin{bmatrix} 3X_1^2 - 2 \\ 2X_2 + 1 \\ -16X_3^3 + 12X_3 \\ +2X_4^{-3} + 5 \end{bmatrix}$$

is the vector that points in the direction of steepest ascent. For any particular point, for example $X_1 = 2$, $X_2 = -4$, $X_3 = 1$ and $X_4 = 1$, the gradient vector is given by

$$\nabla f(2, -4, 1, 1) = \begin{bmatrix} 3X_1^2 - 2 \\ 2X_2 + 1 \\ -16X_3^3 + 12X_3 \\ 2X_4^{-3} + 5 \end{bmatrix} = \begin{bmatrix} 10 \\ -7 \\ -4 \\ 7 \end{bmatrix}$$

Therefore, the gradient vector of $f(X_1, X_2, X_3, X_4)$ at the point 2, -4, 1, 1, is $10X_1 - 7X_2 - 4X_3 + 7X_4$. This vector is in the direction of the maximum rate of change of the function. It should be noted that the gradient is pointing to the steepest path to the ridge of the multidimensional surface but is not pointing to the maximum value of the function. Once the ridge is attained, the gradient evaluated at this point indicates the direction along the ridge toward the maximum value. All this is impossible to visualize in multidimensional space and therefore a two-dimensional problem will be examined to illustrate the technique.

Example 3 Consider the function

$$f(X_1, X_2) = 7X_1 + 4X_2 + X_1X_2 - X_1^2 - X_2^2 = k$$

One must find the values of X_1, X_2 which will maximize $f(X_1, X_2)$. The graph of the function for various values of k are given in Fig. 9.8. The

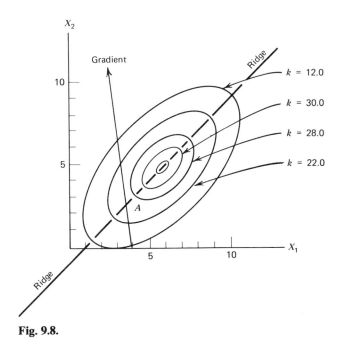

Fig. 9.8.

gradient of the function is given by the vector

$$
\nabla f = \begin{bmatrix} \dfrac{\partial f}{\partial X_1} \\[2mm] \dfrac{\partial f}{\partial X_2} \end{bmatrix} = \begin{bmatrix} 7 + X_2 - 2X_1 \\[1mm] 4 + X_1 - 2X_2 \end{bmatrix}
$$

The direction of the gradient at the point $X_1 = 4$ $X_2 = 0$, which was arbitrarily chosen, is

$$
\nabla f(4,0) = \begin{bmatrix} 7 - 8 \\ 4 + 4 \end{bmatrix} = \begin{bmatrix} -1 \\ 8 \end{bmatrix}
$$

The gradient, also shown in the figure, is pointing the path of steepest ascent and does not point toward the summit. In the ascent, great care must be taken that the distance traversed along the gradient does not go beyond the ridge, point A. At this point ($X_1 \approx 3.7, X_2 \approx 2.8$), the gradient must again be evaluated.

$$f(3.7, 2.8) = \begin{bmatrix} 7 + 2.8 - 7.4 \\ 4 + 3.7 - 5.6 \end{bmatrix} = \begin{bmatrix} 2.4 \\ 2.1 \end{bmatrix}$$

The difficulty is that if too small a step is taken in the direction along the gradient, the convergence process will be slow. If too large a step is taken, instability will occur by going beyond the ridge and starting down the other side of the surface. It was intended to show that the process is not simple and straightforward. A modified technique that is often used is to vary only one component of the vector at a time. This is moderately successful in problems involving only a few variables, but in the average engineering problem involving a large number of variables this technique only complicates matters. It is expected that great progress will be made in the future using these methods, in spite of the complications. Then optimal, rather than feasible, solutions will be obtained for important engineering problems.

References

1. **L. A. Piper and S. A. Hovanessian**, *Matrix Computer Methods in Engineering*, Wiley, 1969.
2. **R. A. Frazer, W. J. Duncan, and A. R. Collar**, *Elementary Matrices*, Cambridge University Press, 1938.
3. **V. N. Faddeeva**, *Computational Methods of Linear Algebra*, Dover, 1959.
4. **H. Lass**, *Vector and Tensor Analysis*, McGraw-Hill, 1950.
5. **J. E. Van Ness**, Response of large power systems to cyclic load variations, *IEEE Trans. PA&S*, Vol. 85, (July 1966), p. 723.
6. **H. Rutishauser**, *Solution of Eigenvalue Problems with the LR-Transformation*, U.S. Department of Commerce, National Bureau of Standards, Applied Mathematics Series 49, *Further Contributions to the Solution of Simultaneous Linear Equations and the Determination of Eigenvalues*.
7. **J. G. F. Francis**, The QR transformation, A unitary analogue to the LR transformation Part I, *Computer J.*, Vol. 4, (October 1961), pp. 265–271; Part 2, *ibid.*, (January 1962), pp. 332–345.
8. **J. A. Pinnello and J. E. Van Ness**, Dynamic response of a large power system to a cyclic load produced by a nuclear accelerator, *IEEE Trans. PA&S*, Vol. 90, (July-August 1971), p. 1857.
9. **C. E. J. Bowler, D. N. Ewart, and C. Concordia**, Self excited torsional frequency oscillations with series capacitors, *IEEE Trans.*, paper T 73-218-5, (January 1973).

10. **J. E. Van Ness, J. M. Boyle, and F. P. Imad,** Sensitivity of large, multiple loop control systems, *IEEE Trans. AC*, Vol. 10, (July 1965), p. 308.

11. **G. B. Dantzig,** Maximization of a linear function of variables subject to linear inequalities, in *Activity Analysis of Production and Allocation,* edited by T. C. Koopman, Wiley, 1951, Chap. 21.

12. **E. M. L. Beale,** *Mathematical Programming in Practice,* Wiley 1968.

13. **P. Wolfe,** The composite simplex algorithm, *SIAM Rev.*, Vol. 7, (1965), pp. 42–54.

14. **J. W. Ballance and S. Goldberg,** Subsynchronous resonance in series compensated transmission lines, *IEEE Winter Power Meeting 1973,* Paper 73-TP167-4-PWR.

15. **M. J. Steinberg and T. H. Smith,** *Economic Loading of Steam Power Plants and Electric Systems,* Wiley, 1943.

16. **E. E. George,** Intrasystem transmission losses. *Trans. AIEE,* Vol. 62, (March 1943), pp. 153.

17. **R. Curant,** *Differential and Integral Calculus,* Vol. 2, Interscience, 1936.

18. **D. J. Wilde and C. S. Brightler,** *Foundations of Optimization,* Prentice-Hall, 1967.

19. **D. A. Pierre,** *Optimization Theory with Applications,* Wiley, 1969.

20. **D. J. Wilde,** *Optimum Seeking Methods,* Prentice-Hall, 1964.

21. **H. W. Kuhn and A. W. Tucker,** Nonlinear Programming, Proceedings of the second Berkley Symposium on Mathematical Statistics and Probability, University of California Press, 1951.

22. **L. Pun,** *Introduction to Optimization Practice,* Wiley, New York, 1969.

23. **D. A. Pierre,** *Optimization Theory and Applications,* Wiley, New York, 1969.

24. **L. K. Kirchmayer,** *Economic Operation of Power Systems,* Wiley, New York, 1958.

25. **J. F. Dopazo et al.,** An optimization technique for real and reactive power, *Proc. IEEE,* Vol. 65, (1967), pp. 1877–1885.

26. **G. Dauphin et al.,** Method of optimizing the production of generating stations of a power network, *Proc. IEEE, 1967 PICA Conf.,* Pittsburg, Pa., pp. 134–140.

27. **R. Bauman,** Power flow solution with optimal reactive flow, *Archiv fur Elektrotechnik,* Vol. 48, No. 4, 1963.

28. **R. Billinton and S. S. Sachdeva,** Optimal real and reactive power operation in a hydro-thermal system, *IEEE Trans.,* 1972.

29. **H. W. Dommel and W. F. Tinney,** Optimal power flow solution *IEEE Trans. PA&S,* Vol. 87, No. 10, (October 1968), pp. 1866–76.

Index

Acceleration of Gauss-Seidel, 107
Adibi, M.M., 203
Anderson, P.M., vii
Anderson, S.W., 176
Andertich, R.G., vii, 5, 122
Angel, R.K., 176

Babic, B.S., 123, 176
Ballance, J.W., 253
Baumann, R., 49, 74, 253
Beals, E.M.L., 253
Bellinton, R., 253
Bolger, R.L., 203
Bonneville Power Administration, 3, 103, 107
Bowler, C.E.J., 252
Boyle, J.M., 253
Bramellor, A., 122
Bree, D.W. Jr., 170, 176
Brightler, C.S., 253
Britton, J.P., 118, 119, 123, 177
Brown, H.E., 5, 48, 74, 101, 122, 125, 176, 203
Brown, R.J., 3, 123
Byerly, R.T., 203
Bus tie Breakers, 46

Caminha, A.C., viii
Carpentier, J., 122, 162, 177
Carter, G.K., 5, 176
Chang, A., 203
Clark, Edith, 100
Coleman, Dorothy, 22, 203
Collar, A.R., 22, 48, 252

Collatz, L., 203
Commonwealth, Edison Co., vii
Concordia, C., 203, 252
Constraints in linear programming, 232
Contingency in power flow, 102, 124-176
 caused by line addition, 138
 evaluation of Newton-Raphson, 125, 159-175
 evaluation by Z-matrix, 129-159
  interchange capability during, 145
  multiple, 135
  single, 129
Coombe, L.W., 5, 23, 48
Customer Loads, 23, 180, 183, 184

Dantzig, G.B., 231, 253
Data for short circuit studies, 28, 50
Dauphin, G., 253
de Mello, F.P., 181
Despotovic, S.T., 123, 125, 176
Determinant evaluation, 13
Diakoptics, 3, 103, 122
Dickson, L., 22
Distribution factors, 125, 175
Dommel, H., 253
Dopazo, J.F., 253
Driving Point Impedance, 4, 24, 26, 29, 30, 38, 45
Duncan, W.J., 22, 48, 252
Dyrkacz, Mary S., 203

Edelmann, H., 5
Eigenvalues and eigenvectors,
    204, 205-231
 applications, 231
 evaluation by
  iteration method, 213
  L.R. transformation, 217
  quadratic convergence, 221
  shifting, 214
El-Abiad, A.H., 48, 101, 125,
    176
Elgerd, D., 203
Equations
 loop, 2, 3, 204
 nodal, 2
Equivalent networks, 34, 49,
    50, 69, 75
 with distribution factors,
    125, 198
 Helmholtz-Thevenin, 26
 reduction of a system to, 59,
    69, 192, 196
Evans, R.D., 100
Ewart, D.W., 252
Exciter response, 188

Faddeava, V.N., 22, 252
Falcone C.A., 176
Fault Current Total, 27, 77,
    99
Feasible solution in linear
    programming, 234, 237
Fortescue, C.L., 75, 100
Francis, J.G.F., 252
Frazer, R.A., 22, 48, 252

Gauss evaluation of a deter-
    minant, 13
Gauss-Seidel Method, 3, 4, 23,
    103, 104-107
 acceleration, 107
Gaussian elimination, 113, 183
Generator reactances, 24, 25,
    182, 183, 184, 185, 197
George, E.E., 253
Glimn, A.F., 3, 5, 123, 177
Goldberg, S., 253
Gradient method of optimiza-
    tion, 249
Gradient vector, 211, 250, 251
Griffin, J.H., 5, 122, 177

Gross, E.T.B., vii

Haberman, R. Jr., 5
Hart, C.E., 5, 122, 177
Hale, H.W., 2, 3, 5, 23, 48,
    102, 114, 159, 166, 176
Happ, H.H., 5, 122, 176
Heydt, G.T., 176
Henderson, J.M., 5
Hovanessian, S.A., 252

Imad, F.P., 253
Iterative methods, see Gauss-
    Seidel Method; Newton
    Raphson
Interchange capability of pool
    operations, 141, 145
Iterative methods instability,
    3, 103

Jacobian Matrix, see Matrix

Kent, M.H., 203
Kimbark, E.W., 180, 202, 203
Kirrchmayer, L.K., 5, 48, 101,
    102, 122, 176, 253
Krou, G., 3, 5, 48, 122
Kron Reduction of a matrix,
    32, 36, 38, 44, 45, 55, 79,
    89, 95, 171, 172
Kruemple, K., 95, 101

Lagrangian Multiplier, 244,
    245
Landgren, G.L., 176
Lantz, M., 5
Lass, H., 252
Lauber, T.S., viii
Lewis, D.G., 5, 23, 48
Limmer, H.L., 125, 176
Line data for short circuit
    studies, 28, 50, 65
Line data reordering, 52
Linear programming, vii, 147,
    204, 231-242
 constraints, 232
 simplex method, 231, 239
Lischer, L.F., vii
Load flow, see Power flow
Lokay, H.E., 203
Long, R.W., 203

Loop equations, 2, 3, 204

Maginniss, F.J., 203
Mastilovic, V.P., 123, 176
Matrix, 1, 2, 6
 adjoint, 11
 Axis discarding, 49, 65, 77,
  125
 cofactor, 10
 definition, 6
 determinant of, 13
 Jacobian, 3, 103, 107, 113,
  117, 159, 160
 identity, 10
 Kron Reduction, 32, 36, 38,
  44, 45, 79, 89, 95, 171,
  172
 major diagonal, 7
 order, 6
 partitioning, 21
 sparcity, 3, 183
 storage, 3
 symmetric, 2, 10
 transpass, 7
 unit, 10
 Y-matrix, 3, 103
 Z-matrix, 3, 4, 24, 47, 49,
  69, 75, 103, 125, 127
  building algorithm, 28, 77
  modified building algorithm,
   58, 81
Matrix algebra, addition, 7
 equality, 7
 identity matrix, 10
 inversion, 11
  classical, 12
  Shipley method, 14
 Kron Reduction, see Kron
  reduction
 multiplication, 8
 subtraction, 8
Matrix types
 connection of network, 2
 incident, 2
Mc Arthur, C.A., 125, 176
Mc Intyre, C.M., 3
Mellor, A.G., 122
Meyer, W.S., 119, 123
Milne, W.E., 202
Mutual Couplings in zero sequence, 80-95

Mutual Couplings in line openings, 95-98

Neid, R.E. Jr., 203
Network, 2
 analyzer, 1, 2, 102
 calculator, 1, 102
 equivalent, see Equivalent
  networks
Network problems, 1, 2, 23, 178
Neuenswander, J.P., 203
Newton-Raphson, 103, 104, 107-
 119, 125, 159, 162
Nodal Equations, 2, 5, 24
Node-definition, 50
Node Voltages, 23, 24

Olive, D.W., 203
O'Mara, J.F., 122
Optimization, vii, 204, 242-252

Person, C.E., vii, 5, 48, 74,
 101, 122, 176
Peterson, N.M., 119, 123, 125,
 170, 176
Pedoe, O., 22
Pierre, D.A., 253
Pinnello, J.A., 252
Piper, L.A., 252
Power Flow, vii, 2, 3, 23, 102,
 203, 205
 Gauss-Seidel, 3, 23, 103, 104-
  107
 high speed, 124-176
 Newton-Raphson, 3, 103, 107-
  118, 159
 Newton-Raphson decoupled, 162
 Z-matrix, 3, 103, 119-122

Quadratic forms, 210

Ralston, A., 177
Reactive power model, 164
Real power model, 166
Reitan, D., 79, 95, 101
Rindt, L.J., 203
Rutishauser, H., 252

Sachdeva, S.S., 253
Santos, J.A. dos R., viii
Sato, N., 5, 107, 122

Saturation, 188, 190
Shipley, R.B., 14, 22, 200,
    203, 241
Short Circuits, vii, 3, 23-100,
    205, 211
Short circuit analysis
 bus tie breakers, 46-47
 line openings, 44, 95
 single phase to ground, 75-100
  coupled lines, 79-100
  line flows, 77, 98
  total fault, 77
 three phase
  line flows, 27, 43
  total fault, 27, 43
 very large systems, 49-74
  by axis discarding, 49, 65,
  77
 using equivalents, 63
Short circuit impedance, 26
Slack variables (linear pro-
    gramming), 233, 239
Smith, T.H., 253
Snyder, W.L. Jr., 176
Stagg, G.H., 3, 5, 48, 74,
    101, 123, 125, 176, 177
State estimation, 204
Steady state stability equiva-
    lent, 196
Steepest ascent, vii, 249, 250
Steinberg, M.J., 253
Stevenson, J.R., 122
Storry, J.O., 101
Stott, B., 119, 123, 125, 176
Symmetrical components, 75, 100

Taylors expansion, 108
Terhune, H.L., 176
Tinney, W.F., 3, 5, 107, 122,
    123, 160, 170, 176, 177,
    181, 253
Torque Accelerating, 178
Transfer impedance, 4, 24, 25,

26, 38, 45
Transformer representation
 Delta-grounded Y, 50
 Ward-equivalent, 166
Transient stability, vii, 178-
    203, 205
 classical representation, 182
 coherent machines, 195
 customers loads, 23, 180, 183,
    184
 dilemma, 181
 exciters and regulators, 188
 saturation effects, 189
 speed versus Details, 181
 transient saliency, 185-188
Transient voltage analysis,
    204
Tucker, A.W., 253

Unit Matrix, see Matrix

Van Ness, J.E., 5, 107, 160,
    177, 252, 253
Venikov, V., 202
Voltage profile, 4, 24, 103,
    105
Voltage regulators, 191

Wagner, C.F., 100
Walker, J.W., 122, 203
Ward, J.B., 2, 3, 5, 23, 48,
    102, 114, 159, 166, 176
Wilde, D.J., 253
Wilf, H., 177
Wolfe, P., 239, 253
 Y-Matrix, see Matrix

Young, C.C., 181, 202, 203

Z-matrix, see Matrix
Zero sequence matrix, see
    Matrix